THE JAPANESE
The Management and Control of an I..................

ABOUT THE AUTHORS

MAX WADE began his careeer in higher education at the University of Wales Institute of Science and Technology, Cardiff, before moving to Loughborough University where he established the International Centre of Landscape Ecology. His interest in invasive alien weeds arises from studies over many years of aquatic plants, a number of key species being Canadian pondweed and water fern. The challenge of managing these non-agricultural weeds brought him out of the water and into contact with plant species such as giant hogweed and Himalayan balsam, as well as Japanese knotweed, enabling him to build up contacts and collaboration across Europe and North America.

LOIS CHILD joined the International Centre of Landscape Ecology in its early days as a research associate, working on a range of projects associated with plant invasions. Over a number of years, she became very involved with the ecology and management of Japanese knotweed, choosing this as the subject for her Ph.D. thesis. In addition to studying these weeds, Max and Lois have been keen to bring experts together to share their knowledge and experience and to run training courses to improve the way we cope with these difficult plants. *The Japanese Knotweed Manual* represents a culmination of these activities.

Max is now working in the Department of Environmental Sciences at the University of Hertfordshire. He has an active involvement in the European Weed Research Society, and he currently chairs the Aquatic Plant Management Group. Lois has been instrumental in establishing the Centre for Environmental Studies at Loughborough University, which she now co-ordinates.

This manual is the first in a series of practical books planned to cover the ecology, management and control of alien and invasive species.

THE JAPANESE KNOTWEED MANUAL

The Management and Control of an Invasive Alien Weed

LOIS CHILD
Centre for Environmental Studies
Loughborough University

MAX WADE
Department of Environmental Sciences
University of Hertfordshire, Hatfield

PACKARD PUBLISHING LIMITED
CHICHESTER

THE JAPANESE KNOTWEED MANUAL
The Management and Control of an Invasive Alien Weed

© L. E. Child and P. M. Wade

First published in 2000 by Packard Publishing Limited,
Forum House, Stirling Road, Chichester, West Sussex, PO19 2EN.

A CIP cataloguing record of this book is available from the British Library.

ISBN 1 85341 127 2

Designed by Cecil Smith.
Typeset in Palatino and Officina by
EverGreen Graphics, Craigweil on Sea, Aldwick, West Sussex.

Printed by Downland Reprographics Limited, West Stoke, Chichester, West Sussex.

Contents

Foreword			x
Acknowledgements			xi
1	**Introduction**		1
2	**The Plant**		4
	Box 2.1	Scientific and common names for Japanese knotweed	5
2.1	**RECOGNITION**		4
	Box 2.2	Key features of Japanese knotweed – mature plant and shoots	6
	Box 2.3	Mean plant height of Japanese knotweed in the East Midlands, United Kingdom, recorded weekly during the growing season (15 April – 3 July, 1994)	7
	Box 2.4	Identification table for Japanese knotweed rhizomes	8
	Box 2.5	Viability test for Japanese knotweed rhizomes	9
2.2	**RELATED SPECIES AND HYBRIDS**		9
2.2.1	**Related species**		9
	Box 2.6	Composition of the genus *Fallopia*	9
2.2.2	**Hybridization**		10
	Box 2.7	Possible routes of hybridization between Japanese knotweed (*Fallopia japonica*) and its related species	11
	Box 2.8	Identification of Asiatic knotweed hybrids	12
	Box 2.9	Comparative leaf shape of Asiatic knotweeds and hybrids	13
	Box 2.10	Distribution of the hybrid *Fallopia* x *bohemica* in the British Isles	14
2.2.3	**Other knotweeds and similar plants**		10
2.3	**HISTORY AND BACKGROUND**		15
2.3.1	**Native habitat**		15
	Box 2.11	Map showing the generalized distribution of Japanese knotweed in Japan	15
	Box 2.12	Varieties of Japanese knotweed in Japan	16
2.3.2	**The spread of Japanese knotweed outside Japan**		17
	Box 2.13	Dates of introduction and naturalization of Japanese knotweed and giant knotweed	17
	Box 2.14	The early history of Japanese knotweed in the UK	18
	Box 2.15	Distribution of Japanese knotweed in a) Europe (1979), b) the British Isles (2000) and c) Cornwall (1999)	18
	Box 2.16	Time-lag between first record of naturalization (1886) and exponential expansion of Japanese knotweed in the UK	20
	Box 2.17	Distribution of Japanese knotweed a) in the USA and Canada and b) in Pennsylvania	21
2.4	**AUTECOLOGY AND HABITATS**		20
2.5	**DISPERSAL AND REGENERATION**		22
2.5.1	**Methods of dispersal and spread**		22

2.6	**IMPACT (PROBLEMS AND LEGISLATION)**	23
2.6.1	**Legal status**	23
	Box 2.18 An example of a legal case concerning Japanese knotweed in the United Kingdom	24
	Box 2.19 Legal status of Japanese knotweed worldwide	25
2.6.2	**Problems caused by Japanese knotweed**	24
	Box 2.20 Problems caused by Japanese knotweed	26
2.6.3	**Costs incurred by Japanese knotweed**	27
	Box 2.21 Summary of comparative treatment costs for development sites contaminated by Japanese knotweed	27
2.6.4	**Advantages of Japanese knotweed**	28
	Box 2.22 Advantages of Japanese knotweed	28
3	**Developing an Effective Programme – Strategic Policy**	29
	Box 3.1 Extracts from the keynote speech at the Japanese knotweed Seminar at Lanhydrock House, Cornwall, 25 November 1997	30
3.1	**RAISING AWARENESS AND PROVISION OF INFORMATION**	29
	Box 3.2 Programme for a Japanese knotweed training workshop	30
	Box 3.3 Sample flier for a talk or lecture	31
	Box 3.4 Websites from the Internet	32
	Box 3.5 A newspaper article drawing attention to a survey of Japanese knotweed	33
	Box 3.6 Outline of main points in designing a poster aimed at helping landowners to dispose of Japanese knotweed safely	34
	Box 3.7 Outline of plan for raising awareness within a local authority	35
3.2	**LIAISON AND CO-ORDINATION**	38
3.3	**ASSESSMENT – SURVEY METHODS, DATA HANDLING, DATA STORAGE, ANALYSIS**	38
3.3.1	**Planning the survey**	38
	Box 3.8 Main stages in undertaking a survey of Japanese knotweed	39
	Box 3.9 Examples of the aims of surveying Japanese knotweed	39
	Box 3.10 Planning a survey	40
3.3.2	**Untertaking the survey**	41
	Box 3.11 Volunteers undertaking a survey of the extent of Japanese knotweed in Swansea, UK	41
3.3.3	**Storing the information**	42
	Box 3.12 a) Record card for storing basic information on Japanese knotweed stands, b) Example of a survey recording sheet	42
	Box 3.13 Distribution of Japanese knotweed stored as a map	44
	Box 3.14 Case study: use of GIS in the City and County of Swansea, UK	46
3.3.4	**Analysis of data**	45
3.3.5	**Further monitoring**	45
3.4	**FORMULATING POLICY**	45
	Box 3.15 Case study: co-ordinated Japanese knotweed management in Cornwall, UK	48
	Box 3.16 Examples of local plans	50

Box 3.17	Extract from an estate management plan of the National Trust	51
Box 3.18	Extract from a river catchment management plan	52
Box 3.19	Extract from City and County of Swansea Policy Document	53

4   Preventing an Invasion   54

Box 4.1	Quotation about avoiding the problems and expense of a full-blown Japanese knotweed invasion	54

4.1   ASSESSMENT OF EXTENT   55

4.2   RAISING AWARENESS AND PROVISION OF INFORMATION   55

4.3   ESTABLISHING A POLICY FOR DEALING WITH THE PREVENTION OF AN INVASION   55

Box 4.2	Containing an invasion of Japanese knotweed in the town of Loughborough, UK	58

4.4   PROGRAMME OF ERADICATION   56

4.5   REASSESSMENT   56

4.5.1   Regular surveys   56

4.5.2   Ongoing surveillance   56

5   Control Options   60

5.1   GENERAL OVERVIEW   60

5.1.1   Choice of methods   60

Box 5.1	Factors to consider when planning a treatment programme	61

5.1.2   Treatment   60

5.2   USE OF EXISTING GUIDELINES   62

Box 5.2	Publications of model specifications for the control of Japanese knotweed in construction and landscape contracts	62
Box 5.3	Extract from the model tender document for eradication of Japanese knotweed	63

5.3   CHEMICAL CONTROL   63

5.3.1   Health and safety   63

Box 5.4	Examples of health and safety legislation in the United Kingdom	64
Box 5.5	Decision-making process prior to selecting a chemical treatment for Japanese knotweed	66

5.3.2   Environmentally sensitive habitats and those near water   65

	Glyphosate	65
Box 5.6	Glyphosate factfile	67
	2,4-D Amine	69
Box 5.7	2,4-D amine factfile	68

5.3.3   Habitats not near water   69

	Picloram	69
Box 5.8	Picloram factfile	70
	Triclopyr	69
Box 5.9	Triclopyr factfile	71
	Imazapyr	69
Box 5.10	Imazapyr factfile	72

5.4   MECHANICAL CONTROL   73

	Box 5.11	Summary of non-chemical control methods	73
5.4.1	**Cutting**		73
	Box 5.12	Cutting: a case study	74
5.4.2	**Mowing**		75
5.5	**MANUAL CONTROL**		75
5.5.1	**Pulling**		76
	Box 5.13	Pulling: a case study	76
5.5.2	**Ineffective control measures**		77
5.6	**BIOLOGICAL CONTROL**		77
5.6.1	**Grazing**		77
5.6.2	**Bio-control agents**		78
	Box 5.14	Comparison of insect herbivore taxa on Japanese knotweed in the UK and Japan	78
5.7	**INTEGRATING CONTROL OPTIONS**		79
5.7.1	**Cutting and spraying**		79
	Box 5.15	Combination treatments – cutting spraying: a case study	80
5.7.2	**Combination of chemical treatments**		79
5.7.3	**Combination of digging and spraying**		79
	Box 5.16	Combination treatments – digging and spraying: a case study	81
5.8	**SITE REVEGETATION**		82

6	**Disposal Options**		**83**
6.1	**DISPOSAL OF SOIL CONTAMINATED WITH RHIZOMES**		83
6.2	**DISPOSAL OF STEM MATERIAL**		83
6.3	**CURRENT GUIDELINES AND LEGISLATION**		84
	Box 6.1	Options for treatment of soil contaminated with Japanese knotweed rhizome fragments	84

| **7** | **Glossary** | | **85** |

8	**Bibliography**		**91**
	8.1	Advantages of Japanese knotweed	91
	8.2	Biological control potential for Japanese knotweed	94
	8.3	Control methods and treatments for Japanese knotweed	96
	8.4	Distribution of Japanese knotweed and related species	101
	8.5	Ecology of Japanese knotweed	105
	8.6	Historical aspects of Japanese knotweed	113
	8.7	Identification of Japanese knotweed	114
	8.8	Problems associated with Japanese knotweed	115
	8.9	Taxonomy of Japanese knotweed and related species	115
	8.10	Other useful resources	117

List of Plates

1   Appearance of Japanese knotweed throughout the year

2   Key features of Japanese knotweed (stem, leaf, flower, rhizome, seed and seed capsule)

3   Japanese knotweed 'crown'

4   Other types of Asiatic knotweed

5   Japanese knotweed in Japan a) on the slopes of Mount Fuji and b) in a lowland location

6   Male and female forms of Japanese knotweed in Japan

7   Japanese knotweed as a primary colonizer on volcanic gravels showing secondary species (*Miscanthus sinensis*) becoming established in the centre of the stand

8   Typical examples of habitats in which Japanese knotweed can be found in its introduced range

9   Japanese knotweed rhizome regeneration

10   Japanese knotweed stem regeneration

11   Japanese knotweed rhizome showing buds and extending shoot

12   A leaflet designed to raise awareness about Japanese knotweed and to provide essential information

13   Poster raising awareness of Japanese knotweed

14   Aerial photograph showing stands of Japanese knotweed

15   Example of GIS output from Swansea survey

16   Appearance of Japanese knotweed one year after treatment with glyphosate

17   Appearance of Japanese knotweed ten days after treatment with triclopyr

Foreword

This manual is based on more than ten years of research by the authors into the life-cycle, biology and ecology of Japanese knotweed. The authors' combined experience in running many training courses and workshops on how to deal with the plant gives the manual a unique insight into the management and control of Japanese knotweed, including raising awareness of its invasive nature and the necessary steps required for preventative action.

Japanese knotweed has become an alien invasive weed throughout Europe, the USA, Canada, and more recently in New Zealand and Australia. This manual provides essential information and guidelines on how to manage this plant, which is causing serious problems in many areas. These range from impeding drainage and displacing native flora and fauna through to damaging buildings and causing blight to sites it has colonized.

The manual is as much to do with preventing an outbreak of Japanese knotweed in an area as it is to do with how to manage the weed once it is established. The text is illustrated with photographs, diagrams and case studies from Europe and North America, and provides an excellent basis for developing a plan either to control the weed or to ensure that it does not become a problem. The manual also provides a valuable template for other invasive species such as giant hogweed and Himalayan balsam.

LOIS CHILD AND MAX WADE
Loughborough and Hatfield
July 2000

Acknowledgements

The manual reflects the work carried out over the last ten years which was variously funded by the Environment Agency (Welsh Region), Welsh Development Agency, City and County of Swansea, Thames Water Utilities plc., National Institute of Agro-Environmental Sciences (Japan) and Leicester and Loughborough Universities Joint Research Fund. This research included inputs from John Bailey (Leicester University), John Brock (Arizona State University), Louise de Waal (University of Wolverhampton), John Palmer and Stephen Blunt (Richards Moorhead and Laing Ltd.).

The authors wish to acknowledge financial support for the preparation of the manual from the Environment Agency (Cornwall Area). Support is also acknowledged from the following people who commented on earlier drafts of the manual: Pip Barrett, Simon Ford (Countryside Manager, North Cornwall, for the National Trust), Sean Hathaway (Japanese Knotweed Officer, City and County of Swansea), Colin Hawke (Cornwall County Council), Deborah Hill (City and County of Swansea), David Holland, Trevor Renals (Biology Team Leader, Environment Agency (Cornwall Region)), Lauren Town (Nomix-Chipman) and Ulla Vogt Andersen (COWIconsult, Denmark).

Thanks are due to Jane Croft (Centre for Ecology and Hydrology, Monks Wood), the Environment Agency, the National Remote Sensing Centre, Ordnance Survey and the Welsh Development Agency for agreement to use illustrations and extracts.

The manual would not have been completed had it not been for the conscientious support from Gill Giles (Loughborough University) and Chris Bedford and Alastair Curry (University of Hertfordshire).

L.E.C. and P.M.W.

1 Introduction

'It is an ill battle where the devil carries the colours.'

Japanese knotweed is a tall, perennial plant with vigorous growth which has been distributed throughout Europe, the United States of America, Canada, New Zealand and parts of Australia. Outside its native range of Japan and northern China, the plant is an aggressive weed which is in an invasive mode. Japanese knotweed is now found along many river and stream corridors, road verges, railway embankments, in gardens and on waste ground. The plant has become a serious weed in many of those areas into which it has invaded and is causing environmental damage and costing many public and private organizations large amounts of money to contain, usually through the application of herbicides.

This manual is aimed at dealing with Japanese knotweed in any part of the world where it has, or might become a problem. A programme for dealing with a Japanese knotweed problem could be considered merely as a control programme: identifying the whereabouts of the plant and taking action against it, for example, by treating with a herbicide. This is a limited approach and is likely to result in only short-term alleviation. A better concept is to use management which, while encompassing control, includes the provision of information about the plant, raises awareness, prevents reinvasion and attracts funding. In order to fit all these elements of management together, a programme needs to be carefully planned. By adopting that approach, a successful outcome can be achieved. This manual seeks to provide the information and guidance for developing just such a programme. The programme could be for a few landowners seeking to remove Japanese knotweed from their land, a local authority working in conjunction with other organizations, a voluntary conservation group or a national agency. Attention should also be given, not only to dealing with an infestation of Japanese knotweed, but towards preventing the plant becoming a problem in the first place. There are still many parts of Europe, North America, Australia and New Zealand, which are presently free of this weed. The manual puts forward the case that both environmentally and economically, it is better to keep the situation that way: 'a stitch in time saves nine'.

The manual begins by considering the recognition and nature of Japanese knotweed (Section 2.1). The nature of the plant is important in understanding the basis of control measures which can be used, such as cutting and the use of herbicides, and emphasis is placed on the roots, or more correctly, the rhizomes. What initially appears to be a simple case of a single problem plant, is a rather more complex story (Section 2.2). Exploring the various types of this plant will not only help to ensure sound management but should also help us to understand more about them and, just as importantly, to avoid even worse problems in the future.

The history of Japanese knotweed and its origins is interesting and also relevant to developing a management programme (Section 2.3). In at least some parts of the world, Japanese knotweed plants have originated from one initial introduction, that is, they are of the same clone. This knowledge certainly makes for more reliable predictions about control, for example, concerning the potential for biological control. In turn, this leads to consideration of the way in which Japanese knotweed spreads and of the problems the plant generates (Sections 2.4 and 2.5). It is also surprising how little legislation or other enforcement there is either to prevent invasion or to enforce control measures (Section 2.6), in view of the likely expenses involved.

This manual is not just about the plant and how to control it. Consideration is given to a number of issues which are of particular importance in ensuring a successful programme of management. These include raising awareness, establishing policy, assessing problems, establishing liaison and co-ordination and appraising the success of any management programme (Section 3). Section 4 goes on to develop a programme to prevent Japanese knotweed from becoming a problem. It is hoped that the manual will be used not only to deal with an existing infestation, but will be used to prevent a problem occurring in the first place.

It is not until Section 5 is reached that control options are described, indicating the need for careful planning and preparation. This section describes chemical, mechanical, and more natural and integrated options for dealing with the plant. Biological control in the sense of using insect (or other invertebrate) herbivores or microbial pathogens such as fungi, while being an attractive option, is not available at the time of writing.

Following mechanical control, there is the problem of how to dispose of the cut or pulled material. This can easily cause reinfestation or disperse the plant to infect another area. Disposal options are considered particularly in relation to soil contaminated by rhizome and cut-stem material in Section 6.

This book is primarily a manual for dealing with a plant which has become a problem from both human and environmental perspectives. It contains a substantial amount of information and indicates the source of other books, scientific papers and reports through a comprehensive list of references. These are ordered by topic (see Bibliography). In order to illustrate the text, a series of examples, case studies and other exemplars are included. These are labelled as Boxes and are referred to in the text. Where appropriate the source of the exemplar is given and/or useful references are included. The manual has been written to be widely accessible and the use of jargon has been kept to a minimum. Any terms which might be difficult to understand have been included in the Glossary, where an explanation can be found.

There is a temptation to develop an almost xenophobic reaction to introduced plants such as Japanese knotweed, and to use terms such as extermination and extinction. It is more important to develop a balanced approach to managing such plants, in providing the relevant information and by building up a database on aspects such as, distribution of the species and costs associated with the problems being caused. The reality is that exterminating this species is unlikely and is, in many cases, unnecessary. There is little international legislation or other enforcement to support the management of such an invasive plant as Japanese knotweed, so much of the programme which is developed in this manual is based on the provision of sound information and advice in order to encourage organizations and individuals to take action against the plant.

While this manual is firmly focused on providing information about Japanese knotweed and how to prevent the plant from becoming a problem, or to deal with the consequences of a full-blown invasion, much of it is applicable to other invasive species. It can be said that in developing policies and an overall programme to deal with these plants, 'He that will not when he may, when he will he shall have nay'.

The Plant

2

'*Bees that have honey in their mouths have stings in their tails.*'

Japanese knotweed is an impressive plant which may already be familiar to some readers. To others it may be a plant heard of, and possibly seen, or about which there is blissful ignorance. Its scientific name is currently *Fallopia japonica*, although it has also been known as *Reynoutria japonica* and *Polygonum cuspidatum* (see Box 2.1). It is also sometimes referred to as *Fallopia japonica* var. *japonica* in order to differentiate it from other varieties of *F. japonica* (see Section 2.2). In Japan, where it originates, it is often referred to as *Polygonum cuspidatum*.

Japanese knotweed produces attractive foliage and creamy white flowers. It is easy to recognize at all stages of growth from young shoot through to mature flowering specimen and as dead stems present during the winter. While there are few other types of plants or genera (see Glossary) with which Japanese knotweed could be confused, there are other species of knotweeds which are similar in appearance. So, regardless of your level of knowledge, you are asked to work through the key features of Japanese knotweed in Section 2.1. The related knotweed species are described in Section 2.2.

2.1 RECOGNITION

Japanese knotweed is a rapidly growing, tall, rhizomatous perennial weed which forms dense thickets. It is easily recognized at all stages of growth. Plate 1 shows the plant at various stages during the year and Plate 2 and Box 2.2 show key features of the mature plant. In spring, the new shoots are green to red/purple in colour and leaves are rolled back. As the shoot extends, the leaves unfurl to show a distinctive flattened base. In its mature state, Japanese knotweed is most easily recognized by its characteristic upright, hollow, bamboo-like stems which are pale green, often with purple speckles, and its arching branches. The leaves form a zig-zag pattern on upper branches which enable the weed to utilize maximum sunlight.

The plant produces small white flowers in elongated clusters arising from the point of the angle between the stem and a leaf. The flowers bloom in late summer, producing a single seed which is enclosed within a three-winged seed capsule (achene). The seeds are triangular in shape about 3 mm (1/8 inch) and dark brown in colour.

The underground rhizome system plays an important part in the life history of Japanese knotweed. Rhizomes (strictly underground stems) act as

Plate 1
Appearance of Japanese knotweed
throughout the year

Spring (right), summer (below)

Plate 1 (cont)
Appearance of Japanese knotweed
throughout the year

Autumn (above), winter (left)

Stem

Plate 2
Key features of
Japanese knotweed

Leaf

Flower

Plate 2 (cont)
Key features of
Japanese knotweed

Rhizome (left)

Seed (3mm) and seed capsule (achene). Reproduced with permission of the publisher, Cornell University Press. Reprinted from Uva, Richard H., Neal, Joseph C. and DiTomaso, Joseph M. *Weeds of the Northeast*. Copyright 1997, Cornell University.

Plate 3 (below)
Japanese knotweed 'crown'

BOX 2.1 Scientific and common names for Japanese knotweed

Japanese knotweed belongs to the plant family Polygonaceae, the knotweeds.
Poly means 'many', gony from the Greek 'knee' meaning jointed.

Fallopia: in honour of G. Fallopia (1523-1562), an Italian anatomist whose name was also given to the Fallopian tubes. *Reynoutria*: in honour of Reynoutre, a 16th century French naturalist.

Scientific name: *Fallopia japonica* (Houtt. Ronse Decraene)
Genus: *Fallopia*
Species: *japonica*
Authority: (Houtt.) Ronse Decraene (The author(s) of the accepted
                     description of the plant, i.e. Houttuyn and Ronse Decraene)

Other scientific names include: *Reynoutria japonica* (Houtt.), *Polygonum cuspidatum* (Siebold and Zuccarini), *Polygonum sieboldii, Polygonum japonicum, Polygonum zuccharini* Small, *Pleuropterus zuccarinii* Small, *Polygonum reynoutria* B & B (in USA horticulture trade).

Common names for *Fallopia japonica*:

Czech Republic:	Kridlatka Japonska
Denmark:	Japan-pileurt
France:	Renouée du Japon, Renouée géantes du Japon
Germany:	Japan-Knöterich, Japanisher Standen Knöterich
Ireland:	Glúineach bhiorach, Glúineach sheapanach
Japan:	itadori (meaning strong plant)
New Zealand:	Asiatic knotweed
Poland:	Rdest Sachalinski
United States of America:	Mexican bamboo, Japanese bamboo, Japanese fleece-flower, wild rhubarb, crimson beauty, elephant-ear bamboo
United Kingdom:	Japanese knotweed, Sally rhubarb, donkey rhubarb, gypsy rhubarb, Hancock's curse
Wales:	Pysen saethwr

Common names for *Fallopia sachalinensis*:

Denmark:	Kæmpe-pileurt
France:	Renouée géantes du Japon
Japan:	o itadori (meaning big strong plant)
New Zealand:	giant knotweed
Poland:	Rdest Sachalinski
South Africa:	sacaline
United States of America:	Sakhalian knotweed, elephant-ear bamboo
United Kingdom:	giant knotweed

BOX 2.2   Key features of Japanese knotweed – mature plant and shoots

Stems
- Annual, develop from rhizome buds at the base of the previous yearís stems in early spring
- Dark red to purple in early spring turning green with vivid red /purple coloured speckles in late spring to summer, becoming deep orange/brown and woody in winter
- Hollow with distinct nodes like bamboo
- Leaves and side shoots arise from nodes
- Semi-woody
- Grow up to 3 m (10 ft) tall in one season, stem diameter may reach up to 40 mm (1.6 in)
- Form dense clumps
- Early frosts can cause wilting of new shoots

Leaves
- Fully grown leaves are heart-shaped, flattened at the base and up to 120 mm (4.7 in) in length
- Young leaves are light green and rolled back, sometimes with red/purple colouring
- Leaves can occasionally show wind damage as pale yellow bands typically one either side and parallel to the midrib.

Japanese knotweed is functionally dioecious, which means that male and female flowers are found on separate plants.

Flowers
- Cream/white coloured
- In clusters 80 to 120 mm (3.2 to 4.7 in) long
- Appear late summer to autumn
- Male flowers erect with protruding stamens
- Female flowers drooping with distinct stigma

NB In some countries both male and female plants are recorded, e.g. in Japan, USA and in Europe, Germany. In others only female plants have so far been recorded, e.g. the United Kingdom and Ireland.

Seeds
- Triangular, 3 mm.(1/8 in) long
- Dark brown and shiny
- Enclosed in thin, papery, three-winged achene

Rhizomes (underground stems)
- Perennial
- Extensive underground system extending up to 7 m (23 ft) away from a parent plant
- Dark brown exterior, 'knotty' in appearance
- Bright orange interior

NB For detailed guidance on rhizome identification see Box 2.4.

storage organs for the plant. Buds are formed in autumn at the base of stems and at nodes along the rhizomes. The plant over-winters in this underground state. In spring, the nutrient reserves stored in the rhizomes are mobilized, enabling rapid growth of shoots from these buds. This rapid increase in early shoot height (see Box 2.3) gives Japanese knotweed a competitive advantage over other species. Regulatory mechanisms ensure that not all buds develop into shoots, thereby conserving resources and regulating shoot density. Generally, buds adjacent to old stems will develop into shoots and over time, large 'crowns' are formed (see Plate 3). During active growth in spring and early summer, the rhizome system provides nutrients for the rapid extension of the stems above ground but once the leaves are fully formed, the flow of nutrients is reversed and photosynthetic products are translocated down to be stored in the rhizomes. The rhizome system may reach a depth of 2 metres (6 feet) and extend up to 7 metres (23 feet) away from a parent plant. Rhizomes range from 5 to 100 mm (¼ to 4 in) in diameter, are fleshy to woody, reddish to dark brown on the outside and a distinctive orange colour inside. In contrast, the true roots of Japanese knotweed are fine, white and thread-like.

It may be necessary to determine whether soil is contaminated with Japanese knotweed in the absence of other parts of the plant, such as stems and leaves. The rhizome identification table in Box 2.4 enables Japanese knotweed rhizomes to be distinguished from other plant roots. If there is uncertainty about the viability of rhizome material, or the presence of

BOX 2.3 Mean plant height of Japanese knotweed in the East Midlands, United Kingdom, recorded weekly during the growing season (15 April – 3 July 1994) (n = 64) (1m = 3.28 feet).   *(Unpublished data of Lois Child)*

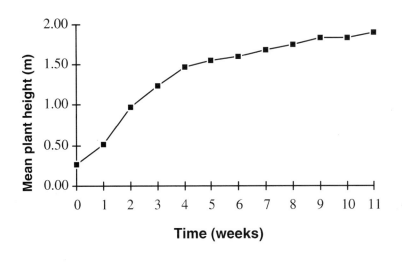

Japanese knotweed needs to be confirmed, a viability test may be carried out (Box 2.5).

Summary

- Japanese knotweed is distinctive in appearance at all stages of growth.
- The underground rhizome system plays an essential role in the life of the plant.

BOX 2.4 Identification table for Japanese knotweed rhizomes

A GENERAL CHARACTERISTICS	YES	NO
Twig-like appearance		
Fleshy with hardness like a carrot		
Brittle when fresh, breaks easily like carrot		
Young rhizomes are white and very soft		

B EXTERIOR OF RHIZOME	YES	NO
Dark brown in colour, like coffee granules		
Texture of outer bark leathery		
When bark is removed the tissue is pale orange/yellow		
Nodes at 1 – 2 cm (0.4 – 0.8 inches) spacing		
Nodes slightly enlarged and knotty		
At nodes white fibrous roots are common		
If present, fresh buds at the nodes are pink in colour		

C INTERIOR OF RHIZOME	YES	NO
Longitudinal view:		
Pale orange to light yellow in colour, carrot coloured		
Central core is dark orange/brown, like rust and sometimes hollow		
Cross section (see Plate 2):		
Cortex with rays coming from pith, giving star shaped appearance		
TOTAL		

If the 'Yes' score is greater than the 'No' score, treat the identified rhizomes as Japanese knotweed.

BOX 2.5 Viability test for Japanese knotweed rhizomes

Rhizome fragments shall be taken from not less than five points in the knotweed area, covering the full extent and depth. Two fragments per point shall be taken, each not less than 150 mm (5.8 in) long and 10 mm (0.5 in) in diameter. Fragments shall be kept damp. Fragments shall be washed to remove soil and placed still wet in a clear polythene bag which shall be loosely tied and labelled with the location of collection point and the date of collection. The bag shall be kept dark at between 10 – 20°C. The bag should be inspected at regular intervals over a 30-day period for signs of root or bud growth. All sample material should be burned at the end of the test.

Source: Welsh Development Agency (1998) *Model Specification for the Control of Japanese Knotweed in Construction and Landscape Contracts.*

2.2 RELATED SPECIES AND HYBRIDS

2.2.1 Related species

The Asiatic knotweed species with which this book deals all belong to the genus *Fallopia* based on the analysis of the plants' chromosomes. A number of these species can be found outside their natural range, for example in Europe. They are typically ornamentals which have escaped to the wild. The genus *Fallopia* includes four sections: *Fallopia, Parogonum, Sarmentosae* and *Reynoutria.* These are summarized in Box 2.6.

BOX 2.6 Composition of the genus *Fallopia*

GENUS: *Fallopia*

SECTION *Fallopia*	SECTION *Parogonum*	SECTION *Sarmentosae*	SECTION *Reynoutria*
annual climbers	perennial climbers	perennial/woody climbers, e.g., Russian vine	herbaceous rhizomatous perennials, e.g., Japanese knotweed and giant knotweed

Giant knotweed (**F. sachalinensis**), native to the island of Sakhalin to the north of Japan, is similar in many respects to Japanese knotweed, but is a much larger plant, 4.0 – 5.0 m (13 to 16 ft) tall. Giant knotweed has leaves which are also much larger (200 – 400 mm, 8 to 16 in long) and are rounded at the base. The plant has greenish flowers carried on shorter, denser panicles than Japanese knotweed (see Plate 4).

A variety of Japanese knotweed, **F. japonica var. compacta**, is more compact in comparison to Japanese knotweed, as its scientific name indicates, usually only reaching 0.7 – 1.0 m (30 – 40 in) tall. It has smaller, more rounded, darker leaves than Japanese knotweed with slightly crinkled edges and can also be distinguished by its slightly reddish-brown flowers and red-tinged stems (see Plate 4).

Russian vine (**F. baldschuanica**), native to Central Asia was introduced to Europe and the USA as an ornamental garden plant. In parts of Europe, it is now rampant on hedges, in thickets and on cliffs in widely scattered localities. Russian vine is a rapidly growing creeper and has been described by the magazine *Gardening Which?* as 'ideal for covering an unsightly shed ... and the house ... the drive ... the car' (see Plate 4).

2.2.2 Hybridization

Hybrid crosses occur between Japanese knotweed and its related species, and Box 2.7 summarizes the various crosses which occur. One of these hybrids, **Fallopia x bohemica**, the hybrid between Japanese knotweed and giant knotweed (*F. japonica* var. *japonica* x *F. sachalinensis*) has become more widely distributed since it was first described in 1983. An illustration of *F. x bohemica* is shown in Plate 4 and a comparison of its characteristics with that of its parental species is shown in Boxes 2.8 and 2.9. As an example, the distribution of *F. x bohemica* in the United Kingdom is shown in Box 2.10. *Fallopia x bohemica* hybrid plants are present throughout Europe and have also been recorded in Australia.

In the United Kingdom at least, all seeds produced on Japanese knotweed plants are hybrid in origin due to the absence of Japanese knotweed male plants (see Section 2.5). The seed is often the product of hybridization with Russian vine (*F. baldschuanica*) although only one mature plant of the hybrid *F. japonica* var. *japonica* x *F. baldschuanica* has ever been recorded. This was a single plant found growing on railway sidings in Haringey, north London in 1992. Seeds produced on Japanese knotweed plants are viable under laboratory conditions but seedlings rarely survive in the wild, possibly due to susceptibility to frosts. There is certainly an absence of data on seedling survival in the wild in Europe and the USA.

2.2.3 Other knotweeds and similar plants

Himalayan knotweed (**Polygonum polystachyum**) can be distinguished from Japanese knotweed by its slightly hairy stems and longer, more slender leaf shape (Plate 4). Growing up to 1.8 m (6 ft) tall, this plant also causes localized problems and is found in similar habitats to Japanese knotweed.

BOX 2.7 Possible routes of hybridization between Japanese knotweed (*Fallopia japonica*) and its related species (with chromosome numbers where known).

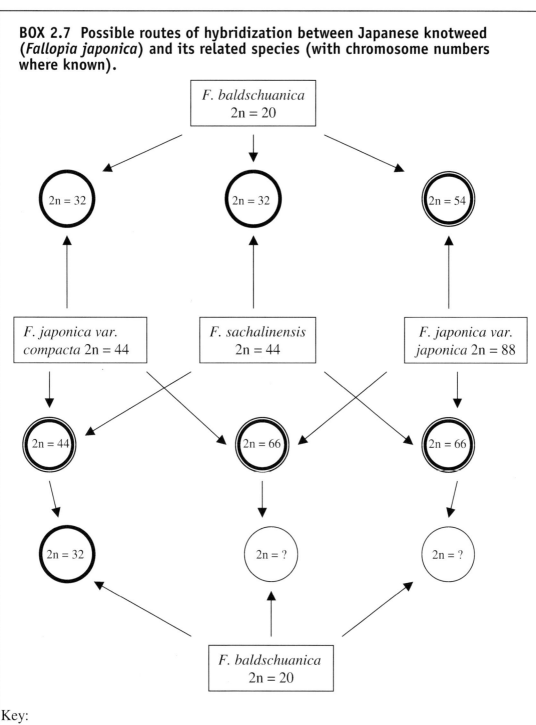

Key:

Found as plants in the British Isles

Not yet found in the British Isles

Found only as seed in the British Isles

Redrawn from: Bailey, J. (1999) The Japanese knotweed invasion of Europe: the potential for further evolution in non-native regions. In: Yano, E., Matsuo, K., Shiyomi, M. & Andow, D.A. (eds), *Biological Invasions of Ecosystems by Pests and Beneficial Organisms. NIAES Series 3*, Tsukuba, Japan.

BOX 2.8 Identification of Asiatic knotweed hybrids

Character	*Fallopia sachalinensis* (Giant knotweed)	*Fallopia x bohemica* (**F. sachalinensis** x **F. japonica** hybrid)	*Fallopia japonica* var. *japonica* (Japanese knotweed)
Chromosome number	2n = 44	2n = 66 or 44	2n = 88
Height	Striking, gigantic plant 4 – 5 m tall (13 – 16 ft)	Habit intermediate 2.5 – 4 m tall (8 – 13 ft)	Large plant to 2 – 3 m tall (6 – 10 ft)
Leaf characteristics	Basal leaves ovate to oblong, base rounded; up to 40 x 22 cm, (15.6 – 8.6 in) length : width ratio *c.* 1.5	Leaves intermediate in size and shape, weakly to moderately rounded at base, tip pointed; up to 23 x 19 cm, (9 – 7.4 in) length : width ratio 1.1 – 1.8	Leaves ovate, tip pointed, base flattened; 10-15 cm long, (3.9 – 5.8 in) length : width ratio 1 – 1.5
	Undersides of leaves with scattered, long, wavy hairs (trichomes)	Undersides of larger leaves with numerous short, stout hairs (trichomes), easily visible with a hand lens	Undersides of leaves without hairs (trichomes)
Sex expression	Male-fertile flowers (with exserted anthers) and male-sterile flowers (with small, empty included anthers and well-developed stigmas) borne on separate plants	Male-fertile and male-sterile flowers borne on separate plants	Flowers usually male-sterile

Other plants with which Japanese knotweed might be confused are:

1. Certain ornamental species with large oval leaves such as *Cornus* species;
2. Large-leaved elm (*Ulmus*) species, the leaves of which are usually noticeably more hairy than Japanese knotweed;
3. Bracken (*Pteridium aquilinum* (L.) Kuhn), the dead fronds of which have a similar colour to the dead stems of Japanese knotweed, especially from a distance, and the summer growth of bracken which may be confused with Japanese knotweed on aerial photographs.

compact variety of Japanese knotweed *(Fallopia*
Reynoutria japonica var. *compacta)* grows to only
1 metre (3 feet) tall, has reddish stems and
more rounded leaves with crinkled edges.

Giant knotweed *(Fallopia* or *Reynoutria sachalinensis)*
is a much taller plant growing to a height of 5 metres
(16 feet). The large leaves have a heart-shaped base.

Plate 4
Other types of Asiatic knotweed

The hybrid between Japanese knotweed
and giant knotweed *(Fallopia x bohemica)*
grows to 2.5 - 3.0 metres (8 - 10 feet) in
height and has leaves slightly larger than
Japanese knotweed with a slightly heart-
shaped base.

Himalayan knotweed *(Polygonum polystachum)* can be distinguished from Japanese knotweed by its slightly hairy stems growing up to 1.8 metres tall and long, pointed leaves.

Plate 4 (cont) Other types of Asiatic knotweed

The Russian vine *(Fallopia baldschuanica)*, although not strictly a knotweed, does form hybrid seeds with Japanese knotweed, and is therefore included in this section. Also known as 'mile-a-minute', it is a fast-growing creeper.

Plate 5

Japanese knotweed in Japan, on the slopes of Mount Fuji (above), and in a lowland location (right)

Plate 6 (left & middle) Male and female forms of Japanese knotweed in Japan

Plate 7 (below) Japanese knotweed as a primary colonizer on volcanic gravels showi secondary species *(Miscanthus sinensis)* becoming established in the centre of the stand, Mount Fuji, Jar Photo: Naoki Adachi

Summary

- There are a number of species closely related to Japanese knotweed which are also ornamental introductions outside their native ranges.
- Japanese knotweed forms hybrid seed with some of these related species.
- Although hybrid seed is formed, in the United Kingdom no germination has been recorded in the wild.

BOX 2.9 Comparative leaf shape of Asiatic knotweeds and hybrids

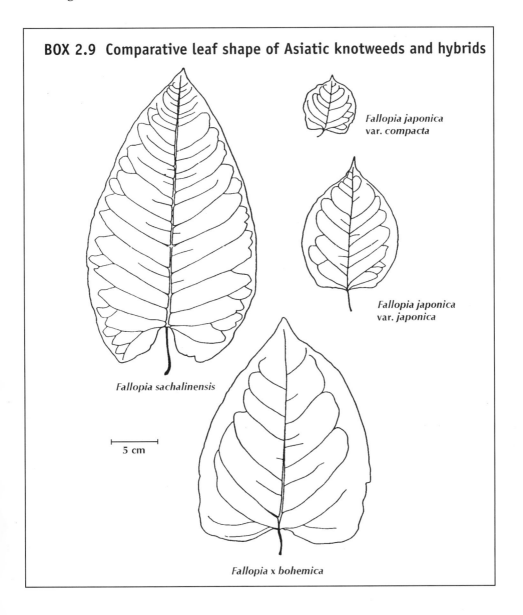

Fallopia japonica var. *compacta*

Fallopia japonica var. *japonica*

Fallopia sachalinensis

5 cm

Fallopia x bohemica

BOX 2.10
Distribution of the hybrid *Fallopia* x *bohemica* in the British Isles

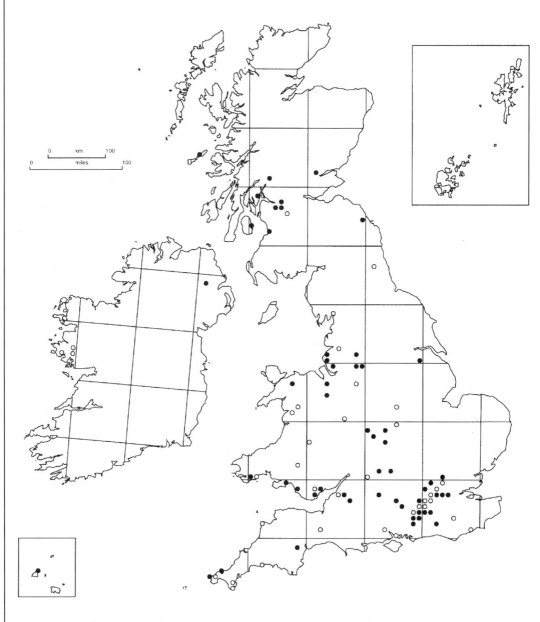

Key: open circles ○ = records prior to 1992; closed circles ● = records 1992 onwards

Source: Centre for Ecology and Hydrology, Monks Wood. 2000. The map was produced using DMAP which was developed by Dr Alan Morton. Each dot represents at least one record in a 10km square of the National Grid.

2.3 HISTORY AND BACKGROUND

2.3.1 Native habitat

Japanese knotweed is native to Japan, Taiwan and Northern China, growing in sunny places in hills and high mountains and along road verges and ditches. Other typical habitats are river gravels, roadsides and managed pastures, especially those where high levels of nitrogen fertilizer are applied. It is found on the islands of Hokkaido, Honshu, Shikoku, Kyushu (see Box 2.11), and in Korea and Formosa. The common name for the plant in Japan is itadori which translates as 'strong plant'. The scientific name often used for Japanese knotweed in Japan is *Polygonum cuspidatum*.

BOX 2.11 Map showing the generalized distribution of Japanese knotweed in Japan

General distribution of Japanese knotweed

The plant can survive harsh environments including acid soils with a pH of less than 4. The soil in which the rhizomes over-winter can be frozen for a number of months. Japanese knotweed also has a high resistance to sulphur dioxide pollution and a low sulphur dioxide absorption rate. It is often found growing alongside volcanic fumaroles with high atmospheric concentrations of sulphur dioxide.

In Japan, the plant forms a natural component of the vegetation and is one of the first species to colonize volcanic lava on which it becomes established (Plate 5). The plant exhibits both male and female forms in its native habitat, that is, separate plants bear male and female flowers (Plate 6). It spreads by seed, which is wind dispersed, and by rhizome fragments, a mechanism of dispersal which enables colonization of newly formed volcanic ash and gravels, in all making an important contribution to the ecosystem development on volcanic gravels such as those found on Mount Fuji. The extensive rhizome system stabilizes the soil, allowing other species to become established, and acts as a reservoir for nitrogen and other essential elements. Japanese knotweed is gradually replaced after approximately 50 years by herb species such as miscanthus grass (*Miscanthus sinensis*) (see Plate 7).

In contrast to the plant in its introduced range, in its native habitats, Japanese knotweed exhibits several forms. A number of varieties are recorded with differences in flower colour, leaf shape and leaf texture (Box 2.12).

**BOX 2.12 Varieties of Japanese knotweed in Japan
(NB In Japan *Fallopia* is described as *Polygonum* or *Reynoutria*)**

Common name	Translation	Scientific name	Description
itadori	strong plant	*Polygonum cuspidatum*	Japanese knotweed
beni itadori	*beni* means red	*Polygonum cuspidatum*	red flower
		P. cuspidatum var. *elata*	pink flower
Hachijo itadori	from Hachijo island	*P. cuspidatum* var. *terminalis*	waxy leaf
		P. cuspidatum var. *compactum*	crinkled leaf edge
ke itadori	*ke* means hairy	*P. cuspidatum* var. *uzenensis*	hairy underside leaf
		P. cuspidatum var. *hastatum*	sword-shaped leaf
o itadori	*o* means big	*P. sachalinensis*	giant knotweed

Summary

- Japanese knotweed is native to Japan, Taiwan and northern China.
- Japanese knotweed is tolerant of a wide range of conditions.
- In its native range the plant is an important component of the vegetation.
- A greater variety of forms of Japanese knotweed exist in its native range compared to its introduced range.

2.3.2 The spread of Japanese knotweed outside Japan

Japanese knotweed was discovered in Japan by Thunberg and was collected in the early to mid-19th century by P. F. von Siebold, an avid collector of Asiatic plants, who sold Japanese knotweed from his nursery in Leiden, The Netherlands. In addition to its value as an ornamental garden plant, Japanese knotweed was sold as a possible species for fixing loose sand and was also recommended as a fodder plant. Specimens of the plant found themselves being distributed around Europe from the mid-19th century onwards (see Box 2.13). Box 2.14 describes in some detail the early history of the plant in the UK. The current distribution of Japanese knotweed in Europe is shown in Box 2.15.

BOX 2.13 Dates of introduction and naturalization of Japanese knotweed and giant knotweed

The date of introduction and naturalization of Japanese knotweed and giant knotweed in those countries for which dates are known:

Country	Japanese knotweed	giant knotweed
Czech Republic (Pysek and Prach, 1993)	1892 introduced	1869 introduced
Europe (Rechniger in Schuldes and Kübler, 1990)	1825 introduced	1869 introduced
Germany (Schemman in: Conolly, 1977)	1823 introduced	1863 introduced
(Alberternst, 1995)	1884 naturalized	
New Zealand	1935 introduced	no data
Russia (Regel in Conolly, 1977)	no data	1864 introduced as a fodder plant
United Kingdom (Conolly, 1977)	1825 introduced	1860 introduced
	1886 naturalized	1896 naturalized
USA (Baily, 1927; Patterson, 1996)	1880s naturalized in north-eastern states	1894 introduced

BOX 2.14 The early history of Japanese knotweed in the United Kingdom

Japanese knotweed was introduced to the UK as an ornamental plant and as a fodder plant.
A specimen was sent to Kew Gardens around 1855 possibly from P. F. von Siebold's nursery in Leiden,
the Netherlands, although the London Horticultural Society apparently obtained a specimen around
1825 (under the scientific name *Houttuynia cordata*). This date is supported by an article in the French
journal *Revue Horticole* (1858) which reports that Japanese knotweed 'has been cultivated for twenty
years in the garden of the Horticultural Society of London'. This specimen apparently came from China
and was planted in an artificial swamp at Kew. Japanese knotweed was soon discovered to survive on
dry ground and was recommended as a plant for the back of the shrubbery or as an isolated specimen
in a part of the lawn. By 1850, the plant had been given the scientific name *Polygonum cuspidatum*
(Siebold and Zuccarini). The invasive nature of Japanese knotweed was soon observed and by 1905 the
Journal of the Royal Horticultural Society no longer advised such planting unless it was 'most carefully
kept in check'.

BOX 2.15 Distribution of Japanese knotweed in a) Europe(1979), b) the British Isles (2000) and c) Cornwall (1999)

a)

Source: Jalas, J. and
Suominen, J. (1979)
Atlas Florae Europaeae.
4. Polygonaceae.
Helsinki.

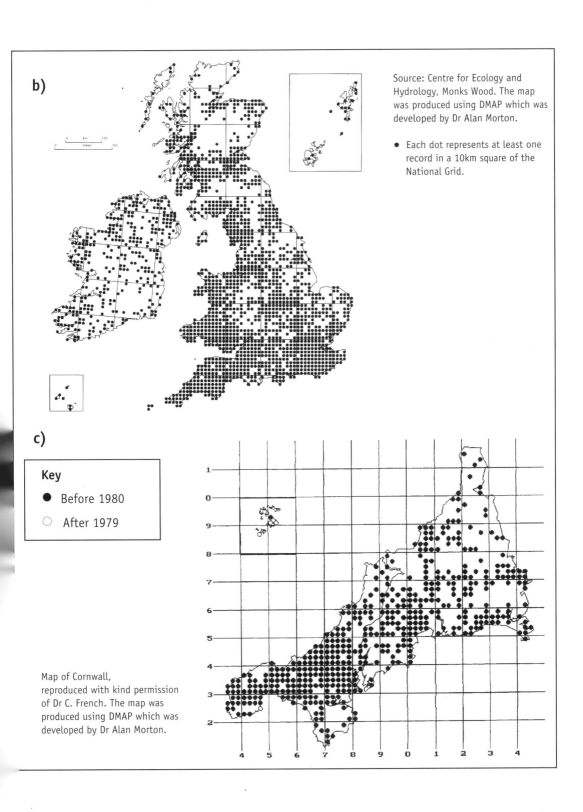

b)

Source: Centre for Ecology and Hydrology, Monks Wood. The map was produced using DMAP which was developed by Dr Alan Morton.

- Each dot represents at least one record in a 10km square of the National Grid.

c)

Key

● Before 1980

○ After 1979

Map of Cornwall, reproduced with kind permission of Dr C. French. The map was produced using DMAP which was developed by Dr Alan Morton.

In the USA, Japanese knotweed was apparently introduced as a garden or yard plant via the British Isles, and as early as the 1880s had become naturalized in north-eastern states. At the turn of the century it was regarded as a plant which could be planted to stabilize soils, especially mine spoil. The means of arrival of the plant in Canada is not clear; likely origins are via the British Isles or with immigrants from Japan and China.

The spread of Japanese knotweed and other related species in Europe was initially quite slow (see Box 2.16) but has become more rapid in the late twentieth century and has now reached pest proportions in many parts of Europe. The current distribution of the plant in North America and Canada is shown in Box 2.17.

Summary

- **Japanese knotweed has expanded its distribution and is invasive in a number of countries outside its native range.**

BOX 2.16 Time-lag between first record of naturalization (1886) and exponential expansion of Japanese knotweed in the United Kingdom

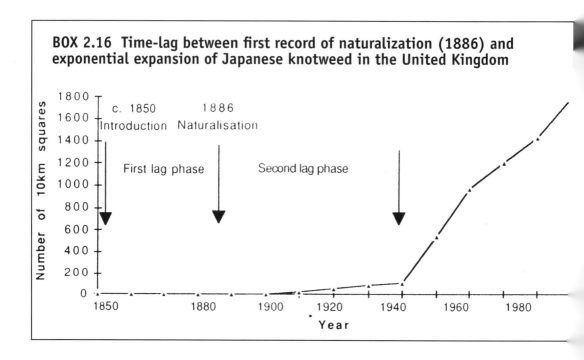

2.4 AUTECOLOGY AND HABITATS

In its introduced range, Japanese knotweed generally occurs in a variety of relatively productive, mostly man-made habitats such as spoil heaps, along canal, stream and river banks, road verges, railway embankments, and various urban habitats (for example, vacant lots, neglected gardens and churchyards) suggesting that human disturbance assists its distribution (Plate 8).

BOX 2.17 Distribution of Japanese knotweed a) in Canada and the USA, and b) in Pennsylvania (PA)

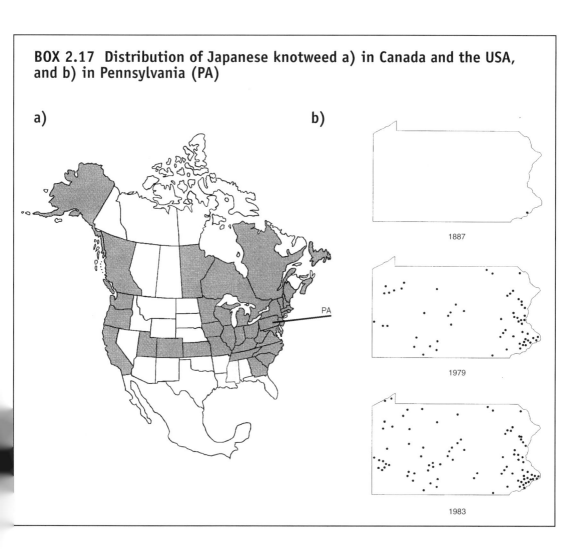

a)   b)

The concentration of Japanese knotweed in urban areas is a reflection of its horticultural origins, the effects of ground disturbance and the less severe late frosts and summer droughts in conurbations to which the shoots are vulnerable.

The success of this species relies on its ability to withstand harsh conditions, such as colonizing cinder tips, railway ballast and other well-drained pioneer habitats in addition to relatively wet areas, such as river banks. The plant tolerates a wide range of pH (3.0 – 8.5) and a variety of soil types. Besides soil acidity, Japanese knotweed shows a remarkable tolerance of habitats polluted by heavy metals or with a high salt content. It is found growing on soils with high levels of heavy metals, such as copper, and low levels of available nitrogen. Plants survive high salinity in coastal areas on beach sands and gravels, cliff faces and on estuarine river banks. Japanese knotweed is not usually found in agricultural land and is not an agricultural weed of any significance.

Despite the occurrence of Japanese knotweed in such a wide variety of habitats, the expansion in riparian areas is the most frequently reported due to the nature conservation, flood defence and recreation problems caused by the plant.

Summary

- **Japanese knotweed is found in many habitats especially along streams and river banks and on disturbed ground.**

- **Japanese knotweed is not commonly recorded as an agricultural weed.**

2.5 DISPERSAL AND REGENERATION

2.5.1 Methods of dispersal and spread

In Japan, sexual reproduction is characterized by high seed production and low seedling survival, with a high probability of survival once seedlings have developed sufficiently deep roots. Although seed is set outside the native range of Japanese knotweed, the role of seeds in the dispersal of the plants is considered to be minimal. In the United States of America, for example, although plants can produce copious amounts of seed, seedling survival in the wild is rare to non-existent. This phenomenon is also reported in the United Kingdom. Much, if not all of this seed source has been found to be of hybrid origin. The presence of fertile male plants of Japanese knotweed is rare in the USA, and in the UK all Japanese knotweed plants recorded to date have been female plants. All male plants in the UK have been shown to be hybrids. No viable seeds of Japanese knotweed have been recorded in the UK except as a result of the hybrid crosses (see Section 2.2). Outside Japan almost all regeneration is therefore exclusively by vegetative means, not only via the extensive and rapidly growing rhizome system (Plate 9) but also from fresh stem material (Plate 10).

With continuing hybridization between the related species of Japanese knotweed, the probability of viable seeds being produced increases. Dependent on seedling survival rates, it may be necessary to programme control strategies in terms of cutting or spraying plants well before flowering in order to prevent seeds being set. Future prevention strategies could include monitoring for seedlings around parent plants in late October and early spring. Results from these studies could then be fed back into control strategies.

The ability of Japanese knotweed to reproduce vegetatively by means of resilient rhizomes, either directly by rhizome extension or by earth cartage containing rhizome fragments, enables large stands to develop. The rhizome system can extend up to 7 m (23 ft) away from a parent plant and down to a depth of 2 m (6.6 ft). Estimates of below ground biomass are given as 14,000 kg/ha dry weight in just the upper 250 mm (9.7 in) of soil. As little as 10 mm length or 0.7 g (0.03 oz) fresh weight of rhizome can give rise to new plants (see Plate 9) and even internode tissue can develop and initiate shoots. In rivers and streams rhizome material washed downstream can regenerate when caught in sediments.

The plant can also be spread by cut-stem material. Under greenhouse conditions, fresh stem segments, 200 – 300 mm (8 – 12 in) long and including two nodes, were placed in water. Shoots and adventitious roots began to be produced from the nodes after just 6 days (see Plate 10). Stems that had produced only shoots in the water treatment after 30 days were subsequently buried in a compost/vermiculite medium and all produced adventitious roots within fourteen days. This indicates that there is the potential for cut stems to regenerate in the wild via dispersal of stem material in rivers and streams. Stems planted in a compost/vermiculite growing medium also gave rise to shoots and roots provided the stems were covered with the planting medium. Once stem material was allowed to dry out, no further regeneration was recorded.

The extensive rhizome system stores carbohydrate over the winter providing a vital source of energy for plant growth the following year. Pinkish nodules develop on rhizomes at the base of stems in the autumn from which the plant produces vigorous shoots in early spring. Rhizome extension is responsible for spread of the plant at a site. The plant may spread at a rate of several metres per year. Long white shoots are produced at rhizome apices and will push up to the soil surface producing new shoots even when buried to a depth of 1m (see Plate 11).

Summary

- **Japanese knotweed regenerates readily from both rhizome and stem material.**

- **The role of male fertile hybrid plants in the spread of Japanese knotweed by seed could add a new dimension to the control measures required for the plant.**

- **Stem material regenerates successfully in both terrestrial and aquatic media.**

2.6 IMPACT (PROBLEMS AND LEGISLATION)

2.6.1 Legal status

Legislation has a role to play in dealing with invasions of plants such as Japanese knotweed. This can be at a number of levels:

1. International legislation restricting the movement of a plant from one country to another;

2. National or state-based legislation preventing movement into or within a country or state;

3. National/state or regional/local measures which require action to be taken against the plant when it poses a problem (an example is given in Box 2.18).

BOX 2.18 An example of a legal case concerning Japanese knotweed in the United Kingdom

**Flanagan vs Wigan Metropolitan Borough Council,
Leigh County Court,
Case Number 93.00392, 26 June 1995**

The above case was brought against Wigan Metropolitan Borough Council by a private landowner whose garden was being invaded by Japanese knotweed from neighbouring Council-owned land. The outcome of the case was that the Council were required to comply with an order to treat a 1 metre (3.3 foot) strip of Japanese knotweed plants along the boundary of the property with glyphosate for a 3-year period. To prevent further spread, the Council was ordered to install a reinforced concrete boundary measuring 0.5 m wide x 10 m long x 1.3 m deep (1.1 ft x 33 ft x 4.3 ft) between the Council-owned land and the private garden to prevent further invasion.

Wigan Metropolitan Borough Council was liable for costs incurred.

Source: Chief Clerk, Court Office, Leigh County Court, 22 Walmesley Road, Leigh, Lancashire, WN7 1YF, UK.

The UK provides a useful example of the types of legislation and control measures which could be used to help in the management of this invasive plant. It is an offence under the *Wildlife and Countryside Act* (1981) to plant or otherwise cause the plant to grow in the wild. In the USA, Japanese knotweed is listed on the National Exotic Pest Plants list and is legally defined as a noxious weed under the Washington State Noxious Weed list. There is currently no legislation which specifically mentions Japanese knotweed in European countries such as Denmark, Germany, Italy, Slovakia or the Czech Republic. A summary of the status of the plant in various countries and states is given in Box 2.19.

See also Section 5.3 for legislation on herbicide use.

2.6.2 Problems caused by Japanese knotweed

Japanese knotweed causes a variety of problems which can be divided usefully into nature conservation, recreation and landscape, flood defence, and the built environment. In many instances, the plant causes more than one problem, for example, alongside a river. Box 2.20, shows a number of situations in which the plant poses problems to environmental managers. Japanese knotweed may also cause damage to archaeological features.

BOX 2.19 Legal status of Japanese knotweed worldwide

LEGISLATION	STATUS OF JAPANESE KNOTWEED
United States	
Eastern Region Invasive Plants	Category 1 – highly invasive
National Exotic Pest Plants list	listed
State of Oregon Noxious Weeds list	List B – Designated weeds
Montana Weed Seed Free Forage Programme	not listed
Brooklyn Botanic Garden Invasive Plants list	not listed
Pacific Northwest Exotic Pest Plant Council status	Red Alert – high potential to spread
State Noxious Weed list (Washington, WAC16-750)	listed and legally defined as a noxious weed
Tennessee Exotic Pest Plant Council (TN EPPC)	listed as one of the most persistent and aggressive of all perennial weeds
State Invasive Exotic Plant Committee (Vermont)	listed as one of 13 most dangerous plants in Vermont
Virginia Natural Heritage Programme	listed as most troublesome alien invasive plant in Virginia
Non-native Invasive Vascular Plants in Connecticut	listed as widespread and invasive (no legal status)
Canada	
Canadian Botanical Conservation Network	invasive potential high
Invasive Alien Plants of Ontario list	especially troublesome
Australia	
New South Wales Noxious Weeds Act	not listed
European Union	no specific legislation
United Kingdom	
Wildlife and Countryside Act 1981	listed under Schedule 9, Section 14 of the Act, it is an offence to plant or otherwise cause the species to grow in the wild
Environmental Protection Act 1990	Japanese knotweed is classed as 'controlled waste' and as such must be disposed of safely at a licensed landfill site according to the EPA (Duty of Care) Regulations 1991

Rio Summit 1992, Convention on Biological Diversity, Article 8(h) calls on participating nations to 'prevent the introduction of, (to) control, or (to) eradiate those alien species which threaten ecosystems, habitats or species'.

BOX 2.20 Problems caused by Japanese knotweed

NATURE CONSERVATION
- Japanese knotweed is a successful competitor. The foliage forms a dense canopy which restricts growth of ground flora and prevents the growth and establishment of other native species.

- Japanese knotweed is able to colonize an area by sending out rhizomes. The plant eventually replaces existing vegetation, including other rhizomatous species such as bracken (*Pteridium aquilinum* (L.) Kuhn).

- The dead foliage stalks persist for 2 – 3 years producing large quantities of debris and slowly decomposing litter which also leads to associated low floristic diversity.

- Sites designated for wildlife interest can be severely damaged.

RECREATION AND LANDSCAPE
- The height to which Japanese knotweed grows can reduce visibility on roadsides and railways.

- Access for anglers and walkers along river banks is made more difficult where there are dense stands.

- In urban areas, stands of Japanese knotweed give an untidy and unkempt appearance and may become used as litter dumps. The dense stems provide a trap for wind blown litter which exacerbates this problem.

- In autumn or winter when the vegetative growth has died back the bare soil exposed is easily washed away, increasing soil erosion, especially on steep river banks.

FLOOD DEFENCE
- In times of flooding, dense stands on the bankside impede flow and may exacerbate flooding.

- Riverbank inspection is made more difficult by the presence of Japanese knotweed in summer.

- Decaying aerial shoots washed into rivers during high winter flows can create instream blockage and increase the risk of flooding, causing damage to urban flood protection schemes.

- Rapidly growing rhizomes and shoots can affect the integrity of flood defence structures by breaking flood revetment blocks apart.

- In autumn or winter when the vegetative growth has died back the bare soil exposed is easily washed away, increasing soil erosion, especially on steep river banks.

MAINTENANCE
- Ground and river bank maintenance costs are greatly increased.

- Maintenance by mowing or flailing can result in further spread of plant by distribution of the plant fragments.

BUILT ENVIRONMENT
- Japanese knotweed shoots are able to push up through asphalt, damaging pavements, car parks and other public facilities including graveyards.

- Rhizomes have been recorded penetrating foundations and other walls, land drainage works and lifting interlocking concrete blocks, causing a wide variety of damage with high associated costs.

- Damage is not only costly to repair, but will become a recurring expense unless the Japanese knotweed is controlled.

BOX 2.21 **Summary of comparative treatment costs for development sites contaminated with Japanese knotweed adjusted for finance costs (i.e. costs incurred in interest payments on land investment), north-east London, United Kingdom**

Treatment	Duration of treatment (months)	Total treatment cost (£/m2) ($/sq yd)	Total finance cost (£/m2) ($/sq yd)	Total cost of treatment adjusted for finance cost (£/m2) ($/sq yd)
Dig and spray	18	1.9	12.0	13.9
(Dig + 2 sprays)		(3.6)	(23.1)	(26.8)
Conventional spray	36	1.3	25.9	27.2
(2 treatments annually for at least 3 years)		(2.6)	(49.8)	(52.4)
(i) Excavate (to 2m depth), cart away and landfill	3	20.9* (40.2)*	1.9 (3.6)	22.8* (43.9)*
(ii) As (i) + import soil and compact	3	49.0* (94.34)*	1.9 (3.6)	50.9* (98.0)*

* excluding Landfill Tax currently at £7.00/cu.m. (*14.74 $/cu. yd.*)
Assumptions: Land values costed at £75/sq.m. (*144 $/sq. yd.*); interest rate of 10%/year rolled up quarterly. Treatment costs from a recent edition of *Spon's Architects' and Builders' Price Book.*

2.6.3 Costs incurred by Japanese knotweed

Control of Japanese knotweed is very costly by manual and/or chemical methods. Treatment has to be continued for several years before effective control is achieved. An example of estimated costs for annual control in a county council in Wales, UK, in 1994 were £300,000 ($500,000).

There are also costs incurred when disposing of soil contaminated with Japanese knotweed rhizomes (see Box 2.21). Disposal of contaminated soil should be by burning or burying, either on site or in landfill according to current guidelines (see Section 6).

2.6.4 Advantages of Japanese knotweed

Some of the advantages of this plant are shown in Box 2.22.

BOX 2.22 Advantages of Japanese knotweed

- It is widely used as a medicine in the Far East, e.g. in China and Korea, for a variety of illnesses;

- It has been used for stabilization of sand dunes and mine spoil, especially in the USA;

- As an ornamental shrub it is a good screening plant with rapid growth, tolerant of a wide range of soil conditions;

- As a source of nectar late in the season for insects such as honey bees;

- As young shoots are edible – historically eaten stir-fried in Japan, with taste a little like rhubarb;

- It is an attractive plant for use in flower arrangements;

- Its stems have been used as a basis for a vegetable dye;

- Dense stands of the plant in urban areas can provide a pseudo-woodland habitat which is of benefit to spring-flowering woodland plants, e.g. bluebell (*Endymion non-scriptus* (L.) Garcke).

Plate 8
Typical examples of habitats in which Japanese knotweed can be found in its introduced range

Plate 8 (cont)
Typical examples of habitats in
which Japanese knotweed can
be found in its introduced range

Plate 9 (above left)
Japanese knotweed rhizome regeneration

Plate 10 (above right)
Japanese knotweed stem regeneration

Plate 11 (left)
Japanese knotweed rhizome showing buds
and extending shoot

Plate 12

A leaflet designed to raise awareness about Japanese knotweed and to provide essential information. Also available on the web: http:www.ex.ac.uk/knotweed/welcome.html

Reproduced by kind permission of Trevor Renals, Environment Agency, on behalf of Japanese Knotweed Control Forum for Cornwall

3 Developing an Effective Programme – Strategic Policy

'Three helping one another, bear the burden of six.'

Developing a programme to deal with a well-established invasive plant will require significant commitment and effort, not just on the part of those intimately involved in the management of the plant but by the other agencies involved, the local community or individuals. Ideally a concerted commitment is needed from all those involved within the area or region. At an early stage, mechanisms should be sought to achieve and cement this commitment. This could be achieved by:

1. Drafting a policy for agreement by the agencies or local community – this must be more than good intentions (see Box 3.1);
2. Establishing a legal basis for the programme, for example, by use of laws from local to national, or even international scales (see Box 2.19)
3. Forming a working group or task force which establishes and agrees terms of reference and a programme of action;
4. Organizing funding.

These mechanisms are not exclusive. For example, the policy established by a local community could include an agreement to establish a Japanese Knotweed Task Force.

3.1 RAISING AWARENESS AND PROVISION OF INFORMATION

Raising awareness and providing information will be important parts of a strategy to deal with Japanese knotweed. Even the prevention of invasion of an area by this plant will require the support of a number of different groups of people. These groups will need to know about the plant and its potential problems with different sub-groups probably being informed at different stages in the overall management programme. For example, drawing the attention of senior staff in an organization, might be necessary early on in order to gain agreement for funding to set up a management programme; while informing members of the public about a menace on their doorstep might best be left till later when the management programme is better developed and their role, if any, is more clearly defined.

BOX 3.1 Extracts from the keynote speech at the Japanese knotweed Seminar at Lanhydrock House, Cornwall, United Kingdom, 25 November 1997

'... I hope that this seminar will be the start of something effective. It is easy to leave an event such as this with good intentions which fail to materialize. It is up to us all to maintain the momentum that this event has generated and translate it into action. As the day progresses, you may become aware of individuals within your own organization that have a role to play. Please approach these people and try to get them on board ...

... There is no single organization which has been given the task of knotweed control. This ambiguity regarding responsibility has been the main reason for our failure to control the spread of knotweed. Whilst we all have more work than we can probably cope with without taking on knotweed control, we cannot continue to deal with this issue in the way which we have. If we continue to ignore the problem. The situation will become unmanageable. There will come a time when we are obliged to act and the opportunity to act effectively will have been missed ...

What today does provide is possibly the last chance we may have for managing knotweed. It is up to us all to use this opportunity to work together and deal with this issue.'

GEOFF BOYD, *Area Manager, Cornwall Area, Environment Agency.*

BOX 3.2 Programme for a Japanese knotweed training workshop

PROGRAMME

9.30	REGISTRATION	12.30	LUNCH
10.00	THE PLANT:	13.30	Legislation
	• Identification	14.00	Best Management Pratice
	• Ecology	15.00	Tea
	• History	15.15	Preventative Measures
10.45	COFFEE	15.45	Long Term Strategy
11.00	The Problems	16.00	Discussion
11.30	Control Methods	17.00	Disperse

BOX 3.3 Sample flier for a talk or lecture

Development of Best Practice
for the Control of

JAPANESE KNOTWEED

25th November 1997
9.00am – 4.30pm
Lanhydrock House, Bodmin

SPONSORED BY THE ENVIRONMENT AGENCY AND
THE NATIONAL TRUST

This event is aimed at anyone involved in the countryside management, conservation, construction and associated sectors who could benefit from a collaborative effort in the control of JAPANESE KNOTWEED

The day will include information on Japanese Knotweed ecology and the latest research from the International Centre of Landscape Ecology, ICOLE, Loughborough and a review of the problem and co-ordinated control approach adopted by Swansea City and Borough. There will be an overview of the problem of Knotweed in Cornwall with talks from the National Trust and Cornwall County Council. Following a buffet lunch two workshops will meet to propose a co-ordinated approach to the problem in Cornwall; one workshop will target the practitioners while the other will aim to develop a co-ordinators group.

(See overleaf for provisional programme and reply slip)

Raising awareness does not have to be expensive. Much can be achieved with a carefully thought-out leaflet targeted at those people who can have an impact on solving the problems caused by Japanese knotweed (see Plate 12). There are various media which can be used in addition to leaflets: poster (see Plate 13), workshop (see Box 3.2), booklet, report, talks and lectures (see Box 3.3), web page on the Internet (see Box 3.4), and via newspaper, radio and television (see Box 3.5).

BOX 3.4 Web sites from the Internet

Invaders Database:
http://invader.dbs.umt.edu/queryplant1.asp

Non-native invasive plant species occurring in Connecticut:
http://darwin.eeb.uconn.edu/ccb/publications/publication-1.html

Pacific North West (PNW) Noxious Weeds:
http://www.tardigrade.org/natives/noxious.html

State of Oregon Noxious Weed List:
http://www.oda.state.or.us/Plant/weed_control/NoxWeedQuar.html#B

NPCI Alien Plant Working Group:
http://www.nps.gov/plants/alien/fact/pocu1.htm

Virtual Gardens:
http://www.rvc.ac.uk/estates/plants/fallopia.htm

Japanese knotweed Control Forum for Cornwall:
http://www.ex.ac.uk/knotweed/welcome.html
http://www.ex.ac.uk/~cnfrench/ics/cbru/monitor/reyjap.htm

Natural Invaders:
http://www.ontarionature.org/enviroandcons/naturalinvaders/invade2.html

University of Missouri:
http://ipm.missouri.edu/ipcm/archives/v7n11/ipmltr5.htm

Tennessee Exotic Pest Plant Council (TN-EPPC):
http://www.webriver.com/tn-eppc/newsletters/v4i3.htm

TN-EPPC Manual:
http://webriver.com/tn-eppc/species/polcus.htm

Technical Information about Japanese knotweed:
http://www.ecy.wa.gov/programs/wq/plants/weeds/aqua015.html

Rutgers Cooperative Extension:
http://rcewebserver.rutgers.edu/weeds/japaneseknotweed.html

Japanese Knotweed Alliance:
http://www.cabi.org/BIOSCIENCE/japanese_knotweed_alliance.htm

Centre for Environmental Studies, Loughborough University, UK:
http://www.lboro.ac.uk/research/cens

BOX 3.5 A newspaper article drawing attention to a survey of Japanese knotweed (*Western Mail*, 13 November 1991)

Fight on to repel Japanese invaders

By DAVID VICKERMAN
Environment Correspondent

AN ALL-OUT assault has been launched in Wales to stem an invasion by Japanese knotweed.

Knotweed "officers" may even be appointed by local authorities as front line troops to tackle the weed, which can spread underground at the rate of 20ft a year and reach seven feet in height.

The autumn offensive comes amid growing concern over the impact of the plant on the native vegetation and landscape of Wales.

Japanese knotweed, introduced to Britain as a garden plant in the 1820s, can grow by more than one foot a week in good weather, shooting up through road surfaces, paving and railway embankments and even sprouting through loose brickwork.

Now the Welsh Development Agency, alarmed at the spread of the plant and the damage caused, has published guidelines on how to spot the plant — and on how to prevent it spreading.

The WDA publications are based on research carried out for the agency by consultants Richards Morehead & Laing, based in Ruthin, Clwyd.

RML was appointed by the WDA in 1989 to examine the extent of the knotweed problem in Wales.

Last night Dr John Palmer, RML's director of environmental sciences, said, "Japanese knotweed is a vigorous and aggressive colonist which can significantly change the landscape and restrict local flora and fauna, as well as causing damage to some man-made structures.

"The first line of defence is for everyone involved in the movement of soil to be aware of the dangers of spreading knotweed by transporting root fragments, and to know what action they should take to prevent that happening."

BOX 3.6 Outline of main points in designing a poster aimed at helping landowners to dispose of Japanese knotweed safely

- Concentrate on particular groups to ensure reaching maximum number of people. There may be a number of very active groups who are influential. Ensure they understand how invasive the plants are and how to deal with cut and dug-up material. Contact local conservation groups known to be concerned about the spread of the plant – ensure they educate all volunteers.

- Make sure that the poster is constructive and informative – avoid scaremongering.

- Emphasize how to dispose of cut or dug-up Japanese knotweed:
    - need to dry, then burn cut stems;
    - potential for composting.

- Provide a list of 'don'ts', e.g. don't tip cut material over the fence.

- Provide list of alternative treatments, e.g. use of herbicides.

- Encourage gardeners to use alternative plant species for screening, etc.

- Summarize contents in a Code of Good Practice – use as a focus for poster.

- List outlets for poster: garden centres, libraries, gardening societies/clubs and allotments, local authority offices.

- Use poster to attract media interest for newspaper or magazine articles, radio and television programmes.

Remember, people have their own priorities and agendas, even in your own organization. Think how best to draw their attention to relevant issues but at the same time avoid upsetting them!

Four examples are provided (Plate 12 and Boxes, 3.5, 3.6 and 3.7) to illustrate this section. The first relates to a leaflet aimed at general awareness-raising by the Japanese knotweed control forum for Cornwall, co-ordinated by the Environment Agency and Cornwall County Council, which was experiencing an invasion of the plant (see Plate 12). Boxes 3.6 and 3.7 provide the notes and key points for helping the general public to deal safely with cut Japanese knotweed (see Box 3.6) and achieving more efficient and effective working within a local authority (see Box 3.7). Box 3.5 is a newspaper cutting.

Raising awareness and providing information should be one of the first tasks in devising a management strategy to deal with Japanese knotweed. Planning an awareness programme should:

1. **Establish the aims of the awareness programme,** for example, in a 'Japanese knotweed free' area. The aims would centre on drawing attention to the undesirable nature of the plant and information to assist its recognition, should the species invade the area, and encourage the reporting of the species to the relevant local authority section.

BOX 3.7
Outline of plan for raising awareness within a local authority

* Decide on the aims of your project for raising awareness.

* Set up a meeting with senior manager(s) to:
    - explain how serious the problem is or could be (show them some figures/examples);
    - ensure that sufficient funding will be available for all aspects of the project and not just weed control;
    - seek to have Japanese knotweed management included in policies of the authority so that it becomes an accepted responsibility;
    - explore the possibility of extending the project to deal with all invasive alien plants.

* Identify whose awareness you plan to raise and whose you would best not involve.

* Determine what information you are going to need in order to achieve your aims.

* How will you actually raise awareness – Use of media? A workshop? Setting up a task group?

* Have you informed relevant individuals and agencies of your plans – Your boss? Other sections in the authority? Agencies you liaise with? Anyone else?

* Check to ensure that funding is available not only for the awareness raising project but for the follow up as well.

* Make plans for the follow-up stage of your project, i.e. after you have achieved your aims for raising awareness.

* Check to make sure that your plan will achieve your aims and make any modifications.

                                                          Good luck!

In an area already experiencing an invasion, the aims could relate to collecting data on the extent of the invasion, seeking funding support, encouraging action to control the plant and/or drawing attention to legislation and preventing further spread.

2. **Determine which groups of people need to be aware of the plant and its problems:** local authority, river management authority, media, local botanists, transport agency, railway company, general public, schools, contractors, and conservation groups. The programme might be directed solely at one group, for example, horticultural suppliers, the aim being to ensure that Japanese knotweed (and other invasive alien plants) are not included in their stock-lists and customers are discouraged from obtaining and planting these species.

3. **Decide what information will be needed.** The information will relate to the aims of the awareness programme and the groups at which it is being targeted. A programme targeted at describing the distribution of Japanese knotweed will need to include such details as good identification information, how to record and report any sightings. The aim might be to encourage liaison between agencies, for example, transport, river and local authorities. Raising awareness of what cost-savings could be achieved by working together and the existing extent of the problem faced by the landowners could be important. A valuable addition to the basic material would be a list of further information, for example, books, addresses, and telephone numbers.

4. **Consider the use of the media:** newspapers, radio and television can be very effective at reaching large numbers of people. A short article in a newspaper could, for example, inform a community that a survey of Japanese knotweed is being undertaken in their neighbourhood (see Box 3.5). A publicity section or officer could assist in constructing and circulating a press release. If not available, take advice or consult a press office to make sure that the press release is newsworthy but avoids any problems.

5. **Achieve liaison with other agencies.** Experience has shown that successfully dealing with Japanese knotweed requires a concerted effort (see Section 3.2). Some agencies will be very aware of the problem whereas others, perhaps those less badly affected, will not. Encouraging a co-ordinated programme of control will require raising awareness amongst all agencies and landowners. This should be aimed at establishing communication and liaison, for example, by a newsletter or an e-mail user group, and/or establishing a working group or task force (see Section 3.2).

6. **Raising funds and dealing with the cost.** Convincing the budget holder or a funding agency to contribute to, or cover, the cost of managing an invasion of Japanese knotweed will often involve persuasion coupled with convincing information. This could be on the basis of financial savings: for example, take action now and avoid spending significant sums

of money later on, or by drawing attention to disruption caused to buildings and paths and/or natural habitats. It may be necessary to work with other agencies, for example, at a regional or national level to lobby for funding (see previous point). Information provision could be through presentations and/or workshops as well as by reports or leaflets.

7. **Consider the implications of the programme.** 'A little information can be a dangerous thing' and a poorly thought-out programme could produce more problems than it solves. Raising awareness can also raise expectations. For example, a local authority project informing the community of the problems and undesirable nature of Japanese knotweed could lead to an expectation that the local authority will deal with the problem. This might not be possible even in the medium to long term. Another response might be for the public taking action, in the absence of follow-up information, such as by cutting down the plant and disposing of the cut material in an inappropriate manner. A programme needs to be thought through from beginning to end, considering all eventualities. This process could be aided by seeking the views of managers, colleagues, the general public and contacts in other organizations.

8. **Prepare for the follow-up from the programme.** This could include:

   - answering enquiries and providing advice;
   - collating data;
   - ensuring support is available, for example, by warning local suppliers of potential increases in demand for glyphosate products;
   - advising colleagues in other sections, other agencies, or on neighbouring property, of potential implications of the programme.

Training has a particular role to play in this process of providing information. This could take a number of forms:

1. Learning about Japanese knotweed (see Box 3.2);

2. Certification in the use of herbicides;

3. Running a workshop on legal aspects, for example, Duty of Care in dealing with soil contaminated with rhizomes;

4. Other aspects, for example, how to write an effective press release and achieve constructive publicity.

A one-day training programme can cover most of the aspects concerning Japanese knotweed. The training programme in Box 3.2 was aimed broadly, but could equally have focused on a specific group, such as local authorities.

Summary
- **Raising awareness is important – 'many hands make light work'.**
- **Establish the aims for an awareness programme carefully:**
  - **different groups of people will need different information.**

- ▪ awareness can be helpful in different ways, for example, recording distribution, fund raising, and control.
- Include the costs associated with an awareness programme in your budget.

3.2 LIAISON AND CO-ORDINATION

There is a temptation to individualize the management of Japanese knotweed and reduce the problem to simply keeping it out of your area. This has been shown to be a minimal approach which inevitably leads to reinvasion and begins a vicious cycle with no real end in sight. A concerted approach to the problem where individuals and organizations work together will be more effective and cheaper.

Liaison could be simply a group of neighbours agreeing to pool their resources and energy and to manage in a co-ordinated manner the Japanese knotweed which has grown up on their land. At a more sophisticated level it could involve establishing a Japanese Knotweed Task Group which includes representatives from a range of organizations which have agreed to work together to deal with the problem over a wide area, like a city or county.

Liaison and co-ordination links closely to the need to provide information and to adopt a policy. The former could take the form of minutes of meetings of the liaison group or a newsletter. A policy could create the framework within which the different members can function with efficiency and minimize duplication, interference and misunderstandings (see Section 3.4).

Summary

- **The persistence, mobility, cost and the fact that Japanese knotweed is no respecter of human boundaries, almost demands that a co-ordinated response is developed with relevant individuals and agencies liaising to address the problem.**
- **Writing and agreeing a policy is essential (see Section 3.4).**

3.3 ASSESSMENT – SURVEY METHODS, DATA HANDLING, DATA STORAGE, ANALYSIS

'Have I got a serious Japanese knotweed problem?' 'Is there any Japanese knotweed in my area?' These are the types of questions which need to be answered early on in order to plan a worthwhile programme. The solution is to undertake a survey. This is essentially a straightforward process but it does need to be carefully thought through. The main stages are summarized in Box 3.8.

3.3.1 Planning the survey

The aims and objectives of a survey (Box 3.8) can vary from the production of a map showing the location of the plant to a map plus associated data about each stand of Japanese knotweed. For example, size, density, land use and proximity to water may be recorded in order to determine factors affect-

---

BOX 3.8
Main stages in undertaking a survey of Japanese knotweed

Develop necessary skills, e.g.

- identification of Japanese
  knotweed and map reading
- ensuring quality control

Plan survey

- define aims
- decide on extent of survey area
- work out timetable
- decide on data storage and analysis

Undertake survey

Store data collected

Analyse data

Decisions

- inform overall programme
- formulation of policy

---

ing the spread and control of the plant. Information could be collected about the past distribution of Japanese knotweed, such as when it was first seen in the area or where the most recently established stands are located. The aim(s) need to be clearly established to make sure the information collected feeds properly into the overall programme.

Closely linked to this will be the extent of the survey area. The assessment can be considered at different scales:

- Continental:
  In which countries or states can the plant be found? Is it still in an expansive phase? (see Box 3.9). Data on the distribution of Japanese knotweed at this scale are poorly collated and governments need to work together more closely to ensure effective surveillance.

BOX 3.9 Examples of the aims of surveying Japanese knotweed

Is Japanese knotweed a potential problem in Anderson County, Cincinnati, Ohio?

What areas of Brittany, France, are free from Japanese knotweed?

Audit the extent of Japanese knotweed along the River Lea corridor, England, as the basis for weed control.

Survey of a Commune, Denmark, in order to determine ownership of land contaminated by Japanese knotweed.

- National or state:
  Many countries or states collect data on the distribution of plant species. These can be funded by governments or states, for example, the Centre for Ecology and Hydrology in the UK, or through non-governmental organizations. The distribution map for Japanese knotweed in the UK in (Box 2.15) shows data recorded where the base unit is 10 km x 10 km squares (6.2 miles x 6.2 miles), that is each black dot represents an area of 100 km^2 (38.4 miles^2). These data are invariably stored electronically on a database and it is relatively easy to generate maps for the distribution of the species at different points in time.

Care needs to be taken in interpreting these maps because:

- records are presented on a presence/absence basis, therefore a single dot could represent a single occurrence of the plant or many hundreds;

- records tend to be made for the first sighting of a species in an area and there are typically few data available to indicate whether the species sustained itself at that location;

- records will not exist for some areas due to lack of surveys, for example, remote areas, though this tends not to be the case for invasive plants such as Japanese knotweed.

BOX 3.10 Planning a survey

- Rather than restricting the survey to within the boundary of an administrative area, include the zone around the boundary or corridors leading into the area, e.g., rivers or roads. This could identify potential sites from which invasion might take place.

- Use existing grid systems as the basis for your recording, such as the National Grid system in the United Kingdom.

- Include training in your planning: a good survey is dependent on skilled surveyors who can not only identify the plant (and differentiate from other similar plants) and can accurately and efficiently collect the other data on the record sheet.

- Plan into your survey good practice in relation to health and safety, e.g. a code of practice for lone working and insurance.

- Contact any landowners for permission to undertake a survey on their land.

BOX 3.11 Volunteers undertaking a survey of the extent of Japanese knotweed in Swansea, United Kingdom

Defining the extent of the study area and the scale of recording, say, 10 km x 10 km or 100 m x 100 m (328 ft x 328 ft), are necessary decisions which govern the type and number of maps needed, size of workforce, time needed to undertake the survey and the cost. A pilot survey of a fraction of the total area would help establish the answers leading to these decisions. Such information is needed even if it is intended to put the survey out to contract.

Deciding on the analysis of the data will influence the survey, for example, in relation to the information to be recorded. A pilot survey would also help to compile or test out a record sheet for the main survey, listing all the data to be collected.

Additional points to consider in planning a survey are listed in Box 3.10.

3.3.2 Undertaking the survey

If a survey has been well planned, implementation should be straightforward. A survey of the city of Swansea in South Wales, United Kingdom, a predominantly urban area with a total area of 96 km² (37 miles²) took approximately 50 person days (see Box 3.11). In contrast, a survey of 200 km² (77 miles²) of the Gower peninsula, South Wales, a rural area, took 60 person days. These surveys were undertaken by fieldworkers using mountain bicycles to improve speed of access around the study area. The former survey took place in late autumn and winter when the dead stems were very obvious, allowing for an easier survey than one undertaken in spring or summer when Japanese knotweed is less obvious in the landscape.

Site monitoring can be conducted either on a visual assessment of percentage cover or a more accurate assessment of stem density can be made by counting stem numbers within randomly placed 1m x 1m (3.3 ft x 3.3 ft) quadrats.

Colour aerial photography can be useful for identifying the location of stands of Japanese knotweed, especially those in less accessible areas, or to estimate the extent of the infestation over a wide area, such as a river catchment. Large stands of the plant have a characteristic form (see Plate 14) but this approach is not detailed enough to pick out stands of only a few or individual plants. Ground survey work can be usefully focused on such nodes which are likely to provide a source of invasion. On aerial photographs, at certain times of the year, Japanese knotweed may be confused with other plants such as bracken (*Pteridium aquilinum* (L.) Kuhn).

3.3.3 Storing information

A systematic and carefully thought-out storage system for the data is essential. Not only do the data need to be accessible and useful for making decisions in the short term, for example in relation to a control programme, but they could also be useful in the future when the control programme is assessed. There are essentially three ways to store such information:

BOX 3.12 (a)
Record card for storing basic information on Japanese knotweed stands

Please read the survey notes on how to record Japanese knotweed	*Survey 2000*
Species:	County:
Grid reference:	Location:
Date:	Name of recorder:
Land ownership:	Habitat:
Identified from:	Abundance:
Dead stem ☐ Flowers ☐	Many dense stands ☐ Many individuals ☐
Green stem/leaves ☐ Rhizome ☐	Few dense stands ☐ Few individuals ☐
Management history:	Additional information:

1. A simple system could be based on record cards or recording sheets which provide details of the location of Japanese knotweed within the study area (see Box 3.12 a & b) and a map of the area showing the location of those plants which were found. The record cards or sheets should include data about the location of the Japanese knotweed together with

---

BOX 3.12 (b) Example of a survey recording sheet

Japanese Knotweed Survey Recording Sheet

International Centre of Landscape Ecology, Department of Geography, Loughborough University, Loughborough, LE11 3TU. Tel: 01509 223030 Fax: 01509 223931

Recorded by		Date	
Site name			
Grid ref.		Site ref.	

Number of stands (**Cut**)			
Total area of cut stands	m		m
Number of stands (**Uncut**)			
Total area of uncut stands	m		m

N.B.   Mark location on the map, as + when stand is <20x20m or linear feature is <25m long, otherwise map as total area.

Average height of stand		<1 m		1-2.5 m		>2.5 m		
Max. stem diameter at 30cm above ground		<1 cm		1-2 cm		2-4 cm		>4cm
Vegetation composition		Japanese knotweed only				mixture of Japanese knotweed and other vegetation		

Proximity to water courses		yes		no		
Slope		flat		moderate		steep

Land Use

	Housing		Shops		Business/Industrial		Public buildings
	Garden		Park		Landscaped area		Recreation grounds
	Farm land		Woodland		Graveyard		Waste ground
	Car park		Road verge		Roundabout		Railway embankment
	River bank		Stream side		Canal		Dock
	Pond		Sea front		Other, specify		

Remarks: ....

BOX 3.13 Distribution of Japanese knotweed stored as a map

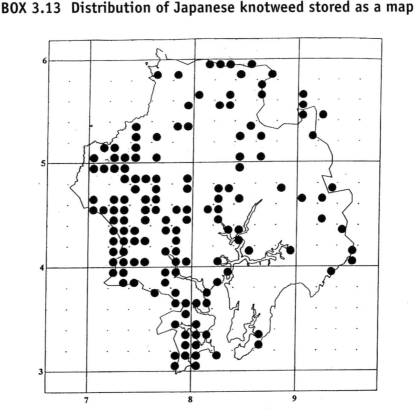

Map of Carrick, Cornwall, 1999, reproduced with kind permission of Dr Colin French. The map was produced using DMAP which was developed by Dr Alan Morton.

the date the data were collected and by whom. Although it may be obvious at the time, the latter items of information could be very useful after a few years when a repeat survey is undertaken. If possible use two maps, one to act as a working map and the other as the 'top' or best copy. Again, this would be valuable in the future.

Cards could be used for the storage of a wider range of data, for example, stand size, soil type, land use and density of plants.

2. The information could be transferred from the survey forms to a computer database. This would typically have the facility of depicting the distribution of the plant as a map which could be used on the computer screen or be printed off (see Box 3.13). An example of a programme which could also be used for this purpose is Recorder, a package produced by English Nature, the government nature conservation body in the United Kingdom.

Japanese knotweed

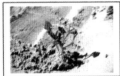

What does it do?

Japanese knotweed
Spreads easily from stems and rhizomes
Grows very rapidly
Invades your garden
Displaces native wildlife
Grows through tarmac
Is expensive to control

DO NOT allow it to spread
DO NOT dig or remove soil within 7 meters of the plant
DO NOT add Japanese knotweed to garden compost

DO take advice on how to treat it
DO report where you have found it to your local authority

Your local authority contact is:
Japanese Knotweed Officer, Nature Conservation Section, Planning Department.

Plate 13 Poster raising awareness of Japanese knotweed

Plate 14 Aerial photograph of Swansea Docks showing stands of Japanese knotweed, top left, and spreading into the sand dunes. Photo courtesy of National Remote Sensing Centre.

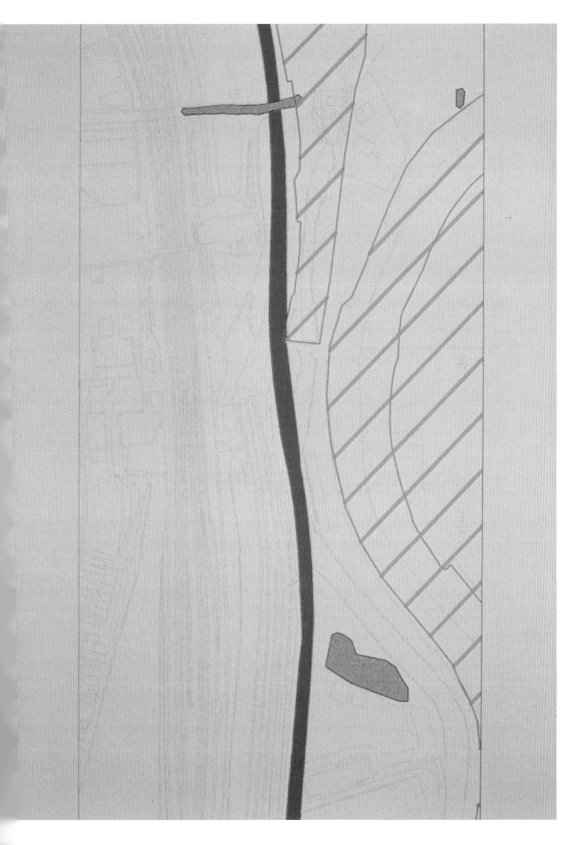

e 15 Example of GIS output from the Swansea survey. The shaded purple areas indicate the presence of nese knotweed; the red area shows proposed highway improvements; the green hatched areas are proposed scape and protection areas. With kind permission of the Ordnance Survey. © Crown Copyright. NC/00/1141

Plate 16 (left)
Appearance of Japanese knotweed one year after treatment with glyphosate

Plate 17 (below)
Appearance of Japanese knotweed ten days after treatment with triclopyr

3. More sophisticated storage of the data could employ a Geographical Information System (GIS). A GIS enables the distribution of Japanese knotweed to be linked with other data, such as land ownership, proposed developments or a stream or a river network. Box 3.14 and Plate 15 provides an example of how a GIS was used in a survey of the city of Swansea, South Wales.

3.3.4 Analysis of data

Making the most of the data is important. A simple distribution map produced by hand can be used to look for patterns such as:

- any corridor effects where the plant is dispersed along a river or a road;
- a concentration of the plant in urban areas as opposed to rural areas;
- Japanese knotweed free areas.

These patterns can be explored more easily using a GIS with a higher degree of reliability and the ability to quantify various components of the data, for example, to assess the percentage area of Japanese knotweed on private land.

GIS could be used to establish an alert system from which a warning would be labelled to any proposed development area containing the plant. This would enable action to be taken to deal with the plant before construction begins and hence could minimize any risk of transporting the plant either around the site or off-site to other areas.

3.3.5 Further monitoring

Repeated site visits will enable maps to be kept up to date and to assess further spread. They will also enable the effectiveness of control methods to be measured. Japanese knotweed free areas must also be checked for any sign of new stands.

Summary

- **A knowledge of the extent of Japanese knotweed within an area is an essential element of developing an effective management plan be it for a garden/yard, a local authority area, a state or a nation. Such information needs to be stored in a useful and accessible form. The data collected need to be analysed to inform the management process and should be updated regularly.**

3.4 FORMULATING POLICY

Establishing and agreeing a policy for dealing with an invasion of Japanese knotweed is essential. This could range from an informal agreement between a group of landowners to contribute jointly to hire a contractor to treat an infestation on and between their properties, through to a policy agreed and published as part of the planning process for a local authority.

BOX 3.14 Case Study:
Use of GIS in the City and County of Swansea, United Kingdom

Aims:
To determine the extent of distribution of Japanese knotweed and to develop a method of storing and retrieving survey data using a Geographical Information System (GIS).

Location:
City of Swansea, South Wales, UK.

Programme of action:
A survey of Japanese knotweed in the City of Swansea was instigated by the City and County of Swansea in 1992. Using 1:2,500 scale Ordnance Survey maps, each occurrence of the plant was recorded directly on both the maps and specially designed recording sheets (see Box 3.12b). Each recording sheet holds a variety of information including a code number for each entry, the site grid reference, the exact area covered by the plant, the associated land use and the proximity of each site to water. The spatial data from the maps were digitized on base maps and the data from the recording sheets were entered on 'Genamap' GIS.

The survey was repeated in 1998 using the same method and data were again transferred to the GIS. The second survey covered an area of 400 km^2 (154 miles^2) which included the area of Swansea city surveyed in 1992 and the surrounding rural area.

Results:
The 1992 survey identified a total area of 48 ha (0.19 miles^2) of Japanese knotweed in the city of Swansea. This amounts to a total area coverage of the city of 0.5%. By 1998 the total had risen to 61 ha (0.23 miles^2) representing a 27% increase in the area covered by Japanese knotweed in the city over the 6-year period. The distribution of the plant through-out the city was uneven with the number of occurrences of the plant within a 1 km x 1 km (0.6 miles x 0.6 miles) square ranging from absent to over 70 locations. The major land use affected by the plant within the city was waste ground (21.7 ha, 0.08 miles^2). Urban areas contributed a significant proportion of land infested (14.1 ha, 0.05 miles^2) and 4.8 ha (0.02 miles^2) of Japanese knotweed was alongside watercourses. Correlating the survey data with land ownership, revealed that approximately 60% of the Japanese knotweed existed on council land, managed by the City and County of Swansea.

Lessons learned:
The advantage of using a GIS for a distribution survey of this type was the ability to link the survey data with other mapped information by overlaying these on the base-line survey. For example, proposals for land redevelopment or road improvements can be linked with the base-line distribution survey, identifying Japanese knotweed-free areas and infested areas (Plate 15). Appropriate control measures can then be brought into action to prevent the plant being spread in the newly developed area by soil disturbance and/or movement. The system, once in place, can be updated at any time to provide an accurate assessment of increased distribution, particular problem areas and effectiveness of treatment.

The database can be used as follows:

- Accurate calculation of the total area covered by the plant within the survey area can be used to assess the treatment programme for the area and associated costs.

- Target areas can easily be identified either as 'priority treatment areas' or as 'protection zones' through the identification of heavily infested areas, intermediate zones and areas free from infestation.

- Base-line data for the monitoring and assessment of further spread of the plant.

- Assessment of the main types of land use associated with the plant in the survey area, which is vital information regarding the choice of treatment method.

- A basis from which to assess the effectiveness of future control measures.

- Identification of major landowners, which can be used in a co-ordinated management strategy.

The production and acceptance of a policy has a number of significant advantages:

- the problem is acknowledged and the need for action agreed;
- an approach to management is agreed with division of responsibilities identified;
- all groups and individuals are made aware of the contents of the policy preventing a fragmented or disparate approach;
- liaison with other organizations is achieved;
- a structured policy may be used to gain funding to support a programme and associated management;
- a policy forms the basis for dealing with other invasive plants or similar problems.

The policy needs to be developed so that it provides a constructive and realistic basis for management. An example of a co-ordinated strategy for management of Japanese knotweed is shown in Box 3.15.

A policy should:
- establish responsibilities and line management;
- set realistic aims and objectives;
- lay down a budget or a plan for achieving funding;
- consider the short, medium and long term, that is, at least five years into the future;
- include means of preventing the spread both within and into the area, as well as controlling existing infestations;
- be reviewed at appropriate times and revised when necessary.

It should not:

- threaten individuals, the community or indeed the organization trying to establish the policy;
- attempt to include too much detail;
- ignore cost implications;
- ignore the need to establish liaison with other agencies and individuals where necessary.

The aims and objectives need to correspond closely to those of the programme being established for Japanese knotweed management. The policy needs to cover all the various stages and incorporate appraisal of progress and a commitment to long-term surveillance for this, and possibly other invasive species.

BOX 3.15 Case Study:
Co-ordinated Japanese knotweed management in Cornwall, United Kingdom

Aims:
To develop best practice for the control of Japanese knotweed.

Location:
Cornwall, UK.

Programme of action:
On 25 November 1997, The Environment Agency, Cornwall Area, and The National Trust, jointly hosted a conference titled 'The development of best practice for the control of Japanese knotweed'. The event was heavily over-subscribed.

The conference provided a good appreciation of the scale of the problem, and how it has been spread. This awareness brought about immediate changes in management practices, such as flailing, which had previously been widespread.

Results:
As a direct result of the conference, the Japanese Knotweed Control Forum for Cornwall (JKCFC) was established. The forum consisted of a co-ordinators group, identified with the strategic management of a co-ordinated control policy, and a practitioners group, developing 'good practice' for the effective control of Japanese knotweed under a range of scenarios.

Each group had representatives from Local Government (District and County level), land-owning industry, estates, trusts and environmental organizations. This created a county-wide network of contacts within organizations with an involvement in knotweed management.

Early efforts were directed toward prevention of spread and public education. A leaflet, 'Japanese knotweed, how to control it and prevent its spread' was produced with

European funds via the Cornwall Landscape Project (see Plate 12). Public interest and support for the initiative was very positive. The initial print run of 15,000 leaflets was distributed within two months. The leaflet proved particularly useful in educating individuals within the haulage, waste and development industries with respect to ease of transmission from soil contaminated with Japanese knotweed material, particularly crown and rhizome.

The publicity campaign also included a recording scheme that was collated and transcribed on to a GIS package by the Botanical Society of the British Isles. This survey provided scope for priority areas for the co-ordinated control campaign. Japanese knotweed growing within areas of high conservation value, areas that had a high potential for infectivity (such as riparian headwaters) and areas with a high risk of imminent disturbance (such as development sites) were identified as priorities.

The group also identified requirements for research, which were largely addressed by the participating organizations. Various methods of control were assessed and collated, and spray trials, using a range of methods and products were compared. Studies were also performed with regard to composting knotweed material.

Lessons learned:
Japanese knotweed management rarely fits into the single remit of any organization. The plant creates many problems on a variety of habitats. It has no respect for boundaries of land ownership. Therefore, in the absence of a passive form of management, such as a biological control agent, effective control can only be achieved by active, co-ordinated co-operation.

To date, a variety of benefits are apparent from the approach adopted in Cornwall. The load has been shared among a variety of organizations and individuals, which has avoided the need to create new posts to address the issue. This approach has also shared ownership for the problem. A network of individuals has formed which can provide advice for specific areas or habitats, depending on the remit of the organization concerned. The co-ordinated approach ensures advice is consistent with our current understanding of 'good practice'.

The level of awareness and knowledge within the local community is impressive. Inappropriate management practices, or the spread of waste and soil containing knotweed, are efficiently reported by the public. The significance of the act is generally well understood by the individuals and companies involved and remedial action is usually prompt and effective.

Members of the group are increasingly being approached by developers to give advice on knotweed management at early stages in planning. A reputation for pragmatic advice has lead to many schemes addressing knotweed problems without significant additional cost or environmental damage. Public awareness regarding the problem and its associated legislation has further encouraged industry and land managers to manage knotweed problems appropriately.

Japanese knotweed control is increasingly being approached in a locally co-operative manner. Agreement between land managers enables control programmes to be performed in the most resource-effective manner, rather than strictly determined by the curtilage of a particular estate or organization.

BOX 3.16 Examples of local plans

The River Erewash is a lowland river in the East Midlands, England, typical of those lowland rivers with both rural and urban sections. The Environment Agency produces catchment management plans known as Local Environment Action Plans (LEAPs). The plan for the Erewash at the consultation stage included a section on 'Eradication of invasive plant species':

'The invasive plant species *Japanese knotweed* is known to be on the River Erewash at Sandiacre and Langley Mill.

This invasive plant species is a problem because it:

- grows extremely densely, shading out native plants;
- provides poor habitat for insects, birds and mammals;
- devalues the natural landscape;
- increases the risk of riverbank erosion when it dies back in the autumn; and
- creates a potential flood hazard if dead stems fall into and clog up watercourses.

The Flood Defence Section has undertaken pilot spraying programmes on the plant at the above sites with some minimal improvements. There needs to be a concerted effort involving the riparian owners to eradicate the plant in the valley to prevent it spreading.'

Source: Environment Agency (1995) *River Erewash – Consultation Report.*

The Wissahickon Creek near Philadelphia is also at an early stage of invasion by Japanese knotweed and includes rural and suburban sections of river. The Wissahickon Watershed Partnership has developed a plan which includes:

- awareness raising through exhibitions, e.g. at main public library;
- information provision via leaflets and advice numbers;
- a programme of treatments, some experimental;
- monitoring of extent of the invasion.

Source: Schuylkill Center, Philadelphia.

Where possible, a policy should be incorporated into existing mechanisms or procedures rather than stand alone as a mere document. This would enable a programme for managing Japanese knotweed to be integrated with other policies and planning. Examples of such documents are:

- local authority plan, such as a Local Plan or a Structure Plan (see Box 3.16);
- an estate management plan (see Box 3.17);
- a river catchment management plan (see Box 3.18);
- national policy on preventing alien species being brought into the country.

BOX 3.17 Extract from an Estate Management Plan of the National Trust (a non-governmental conservation organization) for Rocky Valley, Devon, United Kingdom

ROCKY VALLEY

Sadly Japanese knotweed, *Reynoutria japonica*, is very prevalent along the river bank and has sometimes crossed the footpath or appeared on ledges well above the river. In some areas, this has totally destroyed the ground flora while in others it is just beginning to sprout. There are also large patches of montbretia, *Tritonia* x *crocosmiflora* (crocus), which are spreading and again destroying native flora.

Japanese knotweed and montbretia are a very serious problem in Rocky Valley and require urgent attention.

Management Comments

The most urgent project on the property is to eradicate the Japanese knotweed and montbretia which is beginning to choke the bottom of Rocky Valley. This will be no easy task due to the very difficult terrain but there is the added problem in that there is a ready supply of knotweed further upstream which will continue to colonize the National Trust property if it is not treated in conjunction with Rocky Valley. There is also a substantial quantity on the east side of Rocky Valley although the North Cornwall Heritage Coast and Countryside Service have agreed to assist in eradicating this. A concerted effort by all parties is important in achieving its removal.

Particular monitoring requirements

Map Japanese knotweed and montbretia and ensure that it does not spread to other areas.

Summary of Management Considerations

3.5.2	*LAND MANAGEMENT*	*Date*
3.5.2.3	Eradicate Japanese knotweed and montbretia from Rocky Valley. Work in conjunction with NCHCCS to control opposite bank and upstream.	96/97/ 98/99

Source: The National Trust, *Management Plan: Part A Tintagel Properties*, April 1995 and The National Trust, *Biological Survey, Rocky Valley, Tintagel,* May 1996. Extracted with kind permission of Simon Ford for the National Trust.

BOX 3.18 Extract from a river catchment management plan of the Environment Agency, United Kingdom

3.1 Protection and improvement of our environment

Issue 1: Biodiversity protection

Since the signing of the Biodiversity Convention by the UK Government at the Earth Summit of 1992, biodiversity protection has had a high political profile. Since this time, a National Biodiversity Action Plan has been produced and many county Biodiversity Action Plans are being formulated, including one for Leicestershire. The following are the key sub-issues in the plan area that the Agency is aware of:

Habitats
Alien invasive plant species

There is a need to assess the status and distribution of invasive alien riverside plants (e.g. Japanese knotweed) in the catchment with a view to control or eradication. Such species can have a significant adverse impact on riverine biodiversity.

ISSUE NO: 1		Biodiversity Protection	
OPTIONS/SANCTIONS	*Responsibility*	*Benefits*	*Constraints*
Alien invasive plant species			
i) Survey problem on watercourses	Environment Agency	Gain understanding of the extent of the problem	Survey costs
j) Commence eradication programme as required	Environment Agency	Removal of threat to biodiveristy	Manpower Herbicide use adjacent to water Difficult to eradicate

Source: Environment Agency (1997) *Local Environment Agency Plan, Soar LEAP Consultation Report,* Environment Agency, Nottingham. pp. 27-28.

BOX 3.19
Extract from City and County of Swansea Policy Document

Extract from Japanese knotweed Action Plan for the City and County of Swansea
(adopted June 1997):

Aims:
1. Promote and encourage a co-ordinated response to the control of Japanese
   knotweed in Swansea;

2. Identify priority areas for eradication and establish plans for treatment;

3. Prevent spread into unaffected areas;*

4. Raise awareness and provide information.

*The Action Plan introduced a planning condition regarding likely infestation for all
new planning applications:

Planning officers when visiting sites note the presence of Japanese knotweed
(in addition to using the GIS database) and if confirmed, the following condition is
added to the planning approval decision notice:

> 'full details of a scheme for the eradication and/or control of
> Japanese knotweed shall be submitted to and approved by the Local
> Planning Authority prior to the commencement of work on the site,
> and the approved scheme shall be implemented prior to the use
> of the building/scheme commencing.'

The policy could go so far as to lay down priorities for taking action against Japanese knotweed, for example, in placing an emphasis on preventing invasion of those areas currently free of the plant, setting priorities for areas with high nature conservation value, and tackling upstream sections of water courses in order to remove the potential for spread downstream.

Part of the policy should be aimed at taking action to ensure that the plant is not knowingly brought into the area. This could range from putting Japanese knotweed on a list of plants forbidden to be brought into, for example, a country or state, to passing by-laws prohibiting the importation of soil contaminated by Japanese knotweed rhizomes (see Box 3.19).

Summary

The production and agreement of a policy for dealing with an infestation of Japanese knotweed underpins the whole of the programme of management, assures commitment from those who agree to the policy, and provides structure and guidance.

Preventing an Invasion

4

'A stitch in time saves nine.'

The impression given by distribution maps of Japanese knotweed (see Boxes 2.15 & 2.17) is that there is very little of the land surface which has not been invaded by Japanese knotweed. This is not the case. A dot or shading on the map might represent only one or two plants within a 10 km x 10 km (6.2 miles x 6.2 miles) square (see map of the British Isles in Box 2.15). For example, there are a number of cities and towns in the eastern counties of England, which have not experienced the expansive phase of a Japanese knotweed invasion. However, there are often a few plants within such urban environments which could serve as the basis for a future expansion of the species.

The prevention of an invasion of Japanese knotweed species will take a long time from planning and survey to eradication and re-survey. However, the costs will be minimal compared with having to deal with a full-blown invasion (see Box 4.1). The stages to be included in a preventative programme are:

- assessment of extent of Japanese knotweed in the area;
- awareness raising and provision of information;
- establishing a policy to deal with the prevention of an invasion;
- programme of eradication;
- re-assessment at regular intervals.

BOX 4.1 Quotation about avoiding the problems and expense of a full-blown Japanese knotweed invasion

'One valid criticism for this event would be that it should have happened twenty years ago. With the benefit of hindsight, that might be the case. What today does provide is possibly the last chance we may have for managing Knotweed.'

Source: Geoff Boyd, Area Manager, Cornwall Area, Environment Agency: Keynote speech to Japanese knotweed seminar, 25 November 1997, Lanhydrock House, Cornwall, UK.

4.1 ASSESSMENT OF EXTENT

A survey is needed to determine the extent of Japanese knotweed within the area, or whether the plant has not yet reached the location. The methods which can be used for undertaking such a survey are described in Section 3.3. They include the planning, data collection, storage and analysis. Particular points to note are:

- the density of plants could be relatively low and hence the plant may be difficult to find;
- liaise with botanical societies, wildlife groups and gardeners/horticulturalists as they are likely to have noticed the plant or can keep a look out for it;
- when Japanese knotweed is located look in other likely places, for example, if it is found alongside a stream, look up and downstream from the site and likewise for other corridors such as roads and railways;
- it is important to locate all the stands of Japanese knotweed within the area to ensure as thorough treatment as possible in the control phase;
- use awareness raising to seek support for the survey (see Section 3.1);
- include within the survey a review of the surrounding area, corridors such as rivers running into the area and other nearby towns and cities to determine the nearest stands of the plant.

4.2 RAISING AWARENESS AND PROVISION OF INFORMATION

Raising awareness will be an important part of the programme to prevent an invasion. Not only will it be important to seek help in determining the detailed extent of the plant but it may be necessary to persuade landowners, for example gardeners and private individuals, to control the plant on their land, or at least to ensure that they prevent plants on their land from becoming a source of future invasion. The latter could be particularly difficult if the plant has not become a problem in the area.

Section 3.1 covers this part of the programme.

4.3 ESTABLISHING A POLICY DEALING WITH THE PREVENTION OF AN INVASION

In order to ensure support for a thorough survey, an awareness programme, eradication measures, prevention of future invasion and subsequent resurveys, a policy needs to be formulated and adopted. In the short term this might be necessary to release funds and to encourage agreement that the plant needs controlling even when it is not causing any real problems. In the long term it should include an appraisal of any programme for preventing an invasion including resurveying and, again, the release of funds to ensure that the programme is carried out properly. Consider inviting a person from an area where Japanese knotweed has become a real problem to contribute or advise on establishing a policy.

The policy could also include other undesirable alien species which could become a problem within the area, for example, giant hogweed, purple loosestrife, water fern and Himalayan balsam (see Box 4.2). Part of the policy should be to review the list of such species and to add any other potential invaders from time to time.

Other aspects of establishing a policy are dealt with in Section 3.4 'Formulating policy'.

4.4 PROGRAMME OF ERADICATION

Once the locations of Japanese knotweed are identified, a programme of control can be undertaken. This might be undertaken by, for example, a local authority, a contractor or by different agencies and/or individuals. It is particularly important that all stands are dealt with and if a contractor is used, the contract could be set up to ensure that after a fixed period of time, all the identified stands have been eradicated (see Section 5.2). Some landowners with Japanese knotweed in their gardens may not be prepared to remove the plant. Indeed, Japanese knotweed can have become an established boundary to a garden or yard or an ornamental feature within a border. In such instances, the landowner should be encouraged to take a responsible attitude to the plant and not to allow it to spread. This would include advice on how to dispose of cut or dug material safely (see Section 6.0).

The means of eradication are covered in Section 5.

4.5 REASSESSMENT

Once the Japanese knotweed plants in the area have been dealt with, or it has been ascertained that there is no Japanese knotweed in the area, an agreed programme of re-survey must be established. Two basic strategies are available.

4.5.1 Regular surveys

A structured survey could be undertaken, for example, every three years. A methodology could be set up similar to that used at the beginning of the programme and the re-surveying could be put out to a contractor or ecological consultants. This approach is systematic and should be reliable although early re-invasion could go undetected for three years.

4.5.2 Ongoing surveillance

The surveillance of the plant could be contracted out to local botanists or ecologists either in an informal or formal sense. The plant and other such invasive plants are of sufficient interest that the next occurrence within the area is likely to reported. This approach has the advantage that immediate action could be taken to eradicate any re-invasion. Continual surveys, for example, surveillance in relation to a biological action plan, a river corridor survey, or a botanical survey, could include the recording of alien species as part of their aims. The local community could become involved with groups from local schools, community groups and countryside visitors taking part. As suggested previously, covering a number of undesirable invasive species in the re-survey programme would make good sense.

Summary

The savings, both in terms of environmental damage and money, which could be achieved by preventing an invasion of Japanese knotweed are considerable. There is nevertheless a cost to taking preventative action and a carefully planned programme should include:

- the adoption of policy;
- a survey;
- action to control any stands found;
- follow-up surveillance.

BOX 4.2 Containing an invasion of Japanese knotweed in the town of Loughborough, United Kingdom

Aims:
To deal effectively with the invasion of an urban area by Japanese knotweed.

Location:
Loughborough, UK.

Programme of action:
Loughborough is a town with a population of approximately 50,000. Japanese knotweed is known from only 14 stands or clumps. Examination of the location of these stands indicates that the plant is:

- located in several areas of the town;
- in the process of spreading along at least two corridors;
- in a position to invade a range of sites currently free of the species.

Invasion is at an early stage and there is the opportunity to contain it. This needs to include raising awareness locally about the plant including the intention to eradicate it from the town and the need to establish surveillance for the weed after control. Even though only found at 14 sites, a number of different organizations are involved. The ownership/management of the stands is as follows: 1, the Charnwood Golf Club (a private organization); 2 – 4 grow along the roadside managed by the local authority (Charnwood Borough Council); 5 grows alongside a road managed by the Leicestershire County Council; 6, the Scouts Association (a private organization); 7, Loughborough University; 8 – 10 grow alongside a path maintained by Charnwood Borough Council and a stream, the Wood Brook, managed by the Environment Agency (a national agency); 11 – 13 grow alongside a mainline railway maintained by Railtrack, and 14 grows beside the Loughborough Canal which is maintained by British Waterways (a national agency).

The plant has been spreading along the roads to the west of the town (1 – 5) and along the Wood Brook (8 – 10). The Loughborough Canal site (14) is a new one. Sites 1 – 5 should receive priority action with triclopyr and those along the stream and canal (8 – 10 and 14) need treating with glyphosate. Stands 11 – 13 have been contained by Railtrack for a number of years but the treatments applied have done little to eradicate the stands. In the case of stand 6, it needs to be remembered when planning and implementing treatment that the site is part of a play area for young people. A co-ordinated response could achieve economies of scale if a single contractor were to be used with the contract being carefully written to ensure effective control.

Arising from the liaison betwen the various agencies, a surveillance programme should be agreed.The town has little derelict or waste ground and is located on the edge of the Charnwood Forest, an area of particularly attractive countryside. No major rivers run through the town although the River Soar and the Grand Union Canal run to the north. A number of small infestations of Japanese knotweed were identified throughout the town (see map). In order to establish a comprehensive treatment programme, the landowners were identified and notified of the presence of the plant. This was achieved through media coverage. Some areas of the plant were treated with herbicide.

Results:
Spraying with a selective herbicide was undertaken on the County Council Highways Department land. This achieved good effect for one year, however, new shoots were noticed the following year.

Lessons learned:
• It is necessary to liaise with landowners.
• It is also necessary to maintain a treatment programme for the plant over a number of years.
• Early treatment of plants along watercourses from upstream must be undertaken to avoid further infestations.

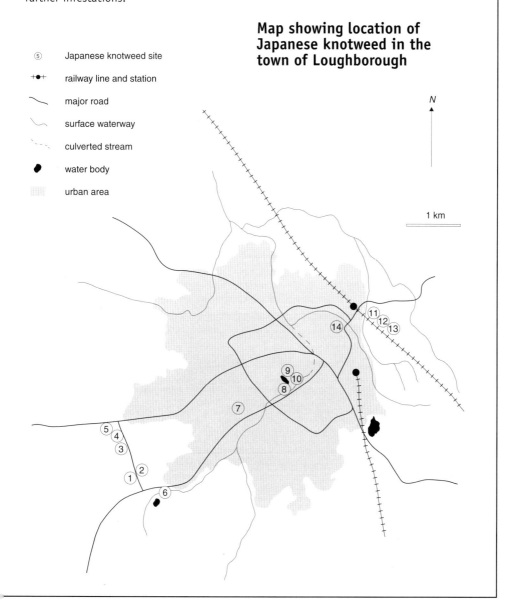

Map showing location of Japanese knotweed in the town of Loughborough

⑤   Japanese knotweed site

╌●╌   railway line and station

‿   major road

‿   surface waterway

╌╌   culverted stream

●   water body

▦   urban area

N

1 km

Control Options

5

'A dead dog has no bite.'

5.1 GENERAL OVERVIEW

5.1.1 Choice of methods

Whatever control method is used, killing the extensive rhizome system is essential if lasting control of Japanese knotweed is to be achieved. This can necessitate a management programme lasting a number of years. It is important to realize that there is no quick solution for the control of Japanese knotweed. Successful control of the plant is achieved largely by chemical application. Although expensive in terms of the chemicals used and the staff hours required to apply them, there are few alternative ways of killing the rhizome system effectively.

The requirements for safety precautions and legislation regarding the use of herbicides differ from country to country. For example in Denmark, no herbicides may be used in or near watercourses. In the USA, individual state laws will govern the use of particular herbicides in specific situations. In the UK, the use of herbicides in water or within 10 metres (30 feet) of a watercourse is restricted and the Ministry of Agriculture Fisheries and Food (MAFF) approves a limited range of products for this use.

Some of the herbicides which are effective against Japanese knotweed are persistent in the soil preventing replanting or reseeding with substitute species. Replanting can be an important part of maintaining control. At sites away from watercourses or drainage channels the status of the land will determine which methods of treatment can be used. In sensitive areas, such as nature reserves and protected sites, mechanical methods of control may have to be employed to prevent damage to other species of identified high nature conservation value. In certain situations, the protection of surrounding vegetation can be achieved by applying herbicide with specialized equipment, for example, a weed-wiper, to target individual Japanese knotweed plants. Some herbicides are selective for broad-leaved plants and can therefore be used in situations where a grass cover is required. Other factors to be considered in the planning of a treatment programme are shown in Box 5.1. In all situations, the control option which achieves the objectives of the control programme and is the least toxic and least damaging to the environment should be chosen.

5.1.2 Treatment

- Chemical treatments should be carried out when the Japanese knotweed is actively growing. In Europe, Japan and North America this is from March to September/October.

- Japanese knotweed can be sprayed with a translocated herbicide in early spring when the plant reaches a height of approximately 1 metre (3 feet) tall. This ensures that there is sufficient leaf area to provide an efficient uptake of herbicide into the leaf and down the stem into the rhizome.

- A herbicide treatment late in the season, before the plant shows signs of senescence, can be very effective. Spraying when the plant is tall may present considerable practical difficulties in terms of access. The key to success at this stage is that the leaves of the plant need to be green to ensure effective uptake of the herbicide (see Section 5.3.1). Spraying with herbicides late in the season when the plant is in flower means that it is necessary to protect bees and other insect visitors to the flowers.

- Cutting, mowing and grazing may commence in early spring and continue through to late autumn (see Box 5.11). Cut material must be disposed of safely (see Sections 2.5.1 and 6.2).

- Manual treatments such as cutting dead stems, pulling and digging, can be carried out during the winter in preparation for subsequent herbicide treatment (see Section 5.5.1 and Box 5.11). All Japanese knotweed material must be disposed of responsibly (see Section 2.5.1 and Chapter 6).

BOX 5.1
Factors to consider when planning a treatment programme

- Proximity of Japanese knotweed to water courses and lakes
- Sensitivity of site, e.g. nature reserve
- Sensitivity of other species of vegetation growing in the vicinity
- Size of infestation
- Bees using the flowers as a nectar source in relation to quantity and quality (herbicide contamination) of honey
- Timing of treatment in relation to plant size and development
- Proximity of tree roots when using certain herbicides such as picrolam and triclopyr
- Density of the stand
- Site access for treatment
- Public access to site
- Need for replanting

5.2 USE OF EXISTING GUIDELINES

There are a number of publications which cover some of the aspects of the control and management of Japanese knotweed. These include:

- *Guidance for the use of herbicides on weeds in or near watercourses and lakes.* Produced by MAFF in the UK.

- *The Use of Herbicides to Control Weeds in or near Water.* Produced by the Environment Agency in the in the UK.

- *Kontrolle des Japan-Knöterichs an Fließgewässern.* Produced in Germany for the control of Japanese knotweeds along watercourses.

- *The Control of Japanese Knotweed in Construction and Landscape Contracts – Model Specification* (1998). Produced by the Welsh Development Agency in the UK, contains detailed guidelines for contractors and specifiers on the control of Japanese knotweed and gives detailed advice on choice of treatment methods and methods used to determine successful control (see Box 5.2).

BOX 5.2 Publications of model specifications for the control of Japanese knotweed in construction and landscape contracts

- *The Eradication of Japanese Knotweed – Model Tender Document* (1998). Produced by the Welsh Development Agency in the UK and includes a detailed description of best practice regarding contract specification, bills of quantities, control of works, completion and payment (see Box 5.3).

BOX 5.3
Extract from the Model tender document for eradication of Japanese knotweed

A contract to control Japanese knotweed, as set out in the Welsh Development Agency's documentation, is deemed to be complete when Japanese knotweed at a site has been eradicated to a minimum of 95% of the original area of cover. This is defined as follows:

1. The area of each pegged Japanese knotweed stand shall be measured.

2. Each Japanese knotweed stand or shoot remaining at the start of the growing season* shall be marked around at a 2 m radius to represent the extent of viable rhizomes.

3. The total area of viable rhizomes (counting overlapped areas once) shall be calculated by the Supervising Officer: if the total area of viable rhizomes remaining is greater than 5% of the total area then the Japanese knotweed control will not be judged to be complete.

*31 May or 6 weeks after emergence of Japanese knotweed shoots, which ever is the later.

Source:
Welsh Development Agency (1998) *The Eradication of Japanese Knotweed – Model Tender Document.*

5.3 CHEMICAL CONTROL

5.3.1 Health and safety

There is a substantial amount of legislation concerned with the use of pesticides. This concerns herbicides which might be considered, their use and the competence of the user. Box 5.4 gives examples of the legislation in the UK.

Before chemical treatment can be carried out, the plant material from previous years must be removed. This can be done either using a brush cutter with a circular metal blade or by trashing dead stems. If fresh material is being cut, the cut material should be piled on site to dry out. Alternatively the material can be removed from the site and burned or disposed of in a landfill site according to current guidelines. In some countries, Duty of Care Regulations will apply (Section 2.6).

BOX 5.4 Examples of health and safety legislation in the UK

In the United Kingdom, the relevant regulations are outlined in *The UK Pesticide Guide*, produced annually.

The **Control of Substances Hazardous to Health Regulations** (COSHH) 1988, covers the use of pesticides and emphasizes that the risks associated with the use of any substance hazardous to health must be assessed before it is used and the appropriate measures taken to control the risk. Consideration must be given whether the use of a pesticide is necessary and if so, the product which poses the least risk to humans, animals and the environment must be selected.

The **Control of Pesticides Regulations** come under the **Food and Environment Protection Act** (FEPA) 1985. Under these regulations, the application of any pesticide should be by contractors with a recognized National Proficiency Tests Council (NPTC) Certificate of Competence. This should ensure that the right herbicides are used, spraying equipment is properly maintained and used, that adequate protective clothing is worn, and that equipment is properly cleaned and waste pesticide disposed of at an approved site.
Detailed guidance on compliance with these regulations is given in the joint MAFF/HSE publication *Pesticides: Code of Practice* for the safe use of Pesticides on farms and smallholdings, 1990. (ISBN 011 242892 4).
Before using any product the user must always read the instructions on the approved label.

Users of pesticides must comply with the conditions of approval stated on the product label. It is an offence to use non-approved products or to use approved products in a manner which does not comply with the specific conditions of approval.

Dependent on the formulation, herbicides can be applied:

- with a tractor mounted sprayer for large areas;
- with a knapsack sprayer for small areas;
- with a long lance sprayer – this is useful in areas where the Japanese knotweed has grown too tall for conventional spraying or where it is inaccessible, for example, on steep slopes and river banks;
- with a controlled droplet applicator;
- applied directly to the individual plants using a weed-wiper or herbicide glove.

WARNING

When using herbicides, always read the product label and comply with the label instructions. It can be an offence to use a herbicide in a manner which is not approved on the label.

Careful application of herbicide will be necessary to avoid damage to non-target plants. All spraying should be carried out in dry weather conditions without rain for 6 hours and preferably 24 hours after treatment, or retreatment may be necessary. Care should be taken to avoid spray drift as this is the most common misuse of pesticides. Spraying is not advisable in wind at speeds greater than Force 2 on the Beaufort Scale (3.2 – 6.4 kph, 2 – 4 mph), measured at a height of 10 m (33 ft), or at low wind speeds of Force 0 (<1.9 kph, <1.2 mph) when vapour drift can occur.

If weather conditions do not permit spraying to be carried out before growth has exceeded 1 m (3 feet) tall, the plant may be cut (see Section 5.3.2) and the regrowth treated later in the season.

A typical decision-making process prior to selecting a chemical treatment for Japanese knotweed is shown in Box 5.5.

5.3.2 Environmentally sensitive habitats and those near water

The only herbicides approved by the Pesticides Safety Precautions Scheme (PSPS) in the UK for use near water and proven to achieve control of Japanese knotweed are specific formulations of glyphosate and 2,4-D amine. Both these herbicides are non-persistent though 2,4-D amine is sometimes preferred for its selective action of targeting only broad-leaved species and leaving a grass cover to persist. In other countries it is important to check that the herbicide formulation is approved for the intended use. You may be required to obtain consent to use a herbicide near water from the relevant environment agency. Consent is required from the Environment Agency for the use of herbicides near water in the UK.

Glyphosate

(For a detailed account of glyphosate, see Box 5.6)

The first application, to arrest growth, can be made early in the season (May/June), with the plant at an approximate height of 1 m (3 feet) and leaves fully developed.

A second application can be made while the plant is still showing signs of regrowth later in the season. Treatment late in the season can be very effective. Planting of trees and shrubs can be carried out seven days after application of glyphosate.

The effects of glyphosate are often slow to appear. A period of at least 21 days after spraying should be allowed before any assessment of effective control is undertaken. Often the full effects of glyphosate treatment are not seen until the following growing season. In the early stages of a control programme, Japanese knotweed plants treated with glyphosate may show a characteristic regrowth and appear as small bushy plants with small pointed leaves (see Plate 16).

BOX 5.5 Decision-making process prior to selecting a chemical treatment for Japanese knotweed

| Is the land due for rapid redevelopment? | YES → | Is there space on site for deep burial? | YES → | EXCAVATE AND BURY |

NO (down) · NO (down from deep burial)

| YES → | Is there space on site for surface treatment? | NO → | EXCAVATE AND LANDFILL |

| Is F. japonica growing amongst vegetation to be retained? | YES → | Is there good grass cover present? | YES → | TREAT WITH SELECTIVE HERBICIDE, e.g. 2,4-D amine, |

NO (down from good grass cover)

SPOT TREAT WITH HERBICIDE, e.g. glyphosate

NO (down)

| Is the site to be planted after treatment? | YES → | TREAT WITH NON-PERSISTENT HERBICIDE, e.g. 2,4-D amine, glyphosate |

NO (down)

| Is the site near water? | YES → | TREAT WITH HERBICIDE APPROVED FOR USE NEAR WATER, e.g. glyphosate, 2,4-D amine |

NO (down)

TREAT WITH HERBICIDE, e.g. triclopyr, imazapyr, picloram

BOX 5.6 Glyphosate factfile

NB. When using herbicides, always read the product label and comply with the label instructions. It is an offence to use a herbicide in a manner which is not approved on the label. For requirements specific to your country or state, consult the agency responsible for pesticide regulation. Wear personal protective equipment as recommended by the product label.

Available as:	**SOLUBLE CONCENTRATE** **EMULSIFIABLE CONCENTRATE** **OIL IN WATER CONCENTRATE**	Approved for professional and amateur use depending on the formulation
Action:	A translocated non-selective non-residual phosphonic acid herbicide	
Active ingredient:	Glyphosate acid	
Application rate	5 l ha^{-1} (1,800g active ingredient ha^{-1}) Increased effectiveness has been achieved using low water volume (ULV) (80 l ha^{-1}) via a conventional knapsack sprayer fitted with a ULV nozzle	
Method of application:	**AS A FOLIAR SPRAY**	Conventional knapsack sprayer Telescopic lance sprayer Controlled droplet applicator Tractor-mounted sprayer
	AS A FOLIAR APPLICATION	Weed wiper Weed glove
	AS A SPOT TREATMENT	Drench gun Cut stump treatment
Time of application:	During active growth (April – September) Treat new growth in spring when leaves are fully expanded When plants are > 0.5 < 1.0 m tall (2-3 feet) (larger plants may be treated using a telescopic lance sprayer) Follow up application in late summer	
Pre-treatment:	Cut back or trash dead stems at the end of the season in order to facilitate access (see Section 5.4.1)	
Limitations:	This is a non-selective herbicide which will have an effect on all vegetation including grasses. Not suitable for use where other vegetation is to be retained. Individual plants may be spot-treated using a weed wiper or weed glove. Use spray hoods when spraying in established forestry plantations	
Effects:	The effects of glyphosate take at least 21 days to show. Regrowth from treated plants has a characteristic stunted appearance (see Plate 16)	
Special requirements:	Use the product as specified on the labelTake extreme care to avoid drift. A rain-free period of at least 6 hours (preferably 24 hours) must follow spraying. Wear personal protective equipment as recommended by the product label	
Adjuvants:	Approved adjuvants may be added to improve the efficacy of the treatment, however, some may not be approved for use in or near water	
Country specific information:		
UK:	Glyphosate is approved for use in amenity and forestry situations. Specific formulations of glyphosate are approved for use in or near water, e.g. Roundup Biactive and Roundup Pro. Always obtain consent from the Environment Agency if spraying within 10m of a water course Glyphosate does not persist in the soil and can be used in sensitive areas such as nature reserves; gives total vegetation control so care should be taken in sensitive areas	
Denmark:	Approved for use in non-cultivated areas	

BOX 5.7 2,4-D amine factfile

NB. When using herbicides, always read the product label and comply with the label instructions. It is an offence to use a herbicide in a manner which is not approved on the label. For requirements specific to your country or state, consult the agency responsible for pesticide regulation. Wear personal protective equipment as recommended by the product label.

Available as:	**SOLUBLE CONCENTRATE**	Approved for professional
	EMULSIFIABLE CONCENTRATE	and amateur use depending
	OIL IN WATER CONCENTRATE	on the formulation
	ULTRA LOW VOLUME LIQUID	
Action:	A selective translocated phenoxy herbicide	
Active ingredient:	2,4-D acid equivalent	
Application rate:	2790 g active ingredient ha^{-1}	
Method of application:	**AS A FOLIAR SPRAY**	Conventional knapsack sprayer
		Telescopic lance sprayer
		Controlled droplet applicator
		Tractor-mounted sprayer
	AS A FOLIAR APPLICATION	Weed wiper
		Weed glove
Time of application:	During active growth (April – September)	
	Treat new growth in spring when leaves are fully expanded	
	When plants are > 0.5 <1.0 m (2-3 feet) tall	
	(larger plants may be treated using a telescopic lance sprayer)	
	Follow up application in late summer	
Pre-treatment:	Cut back or trash dead stems at the end of the season to facilitate access (see Section 5.4.1)	
Limitations:	Suitable for use where grasses are to be retained.	
	This is a selective herbicide which will kill most broad-leaved plants and so can be used on grassland.	
	Individual plants may be spot-treated using a weed wiper.	
	Re-establishment of vegetation is possible after treatment but may not be necessary where grasses persist	
Special requirements:	Use the product as specified on the label	
Adjuvants:	Do not spray if rain falling or imminent	
	Wear personal protective equipment as recommended by the product label	
Country specific information:		
UK:	Specific formulations of 2,4-D, such as Dormone, are approved for use in or near water; many formulations are NOT approved for use in or near water, e.g. Agricorn 2,4-D may be used in established grasslands, amenity and forest situations. Always obtain consent from the Environment Agency if spraying within 10 m of a watercourse	
Denmark:	Approved for use in non-cultivated areas	
	No herbicides are approved for use in waters of, e.g. streams and lakes	

2,4-D amine

(For a detailed account of 2,4-D, see Box 5.7)
The first application can be made early in the season (May) to reduce height and vigour of the plant. A second application may be made in mid/end-season. A treatment late in the season can be very effective

When using herbicides, treatment may need to be continued over several years until no new shoots appear.

5.3.3 Habitats not near water

In the UK for example, in habitats where there is no risk of runoff into surface drains and water courses, the following herbicides may be used. In other countries, check that the herbicide formulation is approved for the intended use. Care should be taken to ensure that other plants, especially trees are not affected if roots run under or near to treated areas.

Picloram

(For a detailed account of picloram, see Box 5.8)
Single foliar application during the growing season when plants have reached a height of one metre (three feet).

Triclopyr

(For a detailed account of triclopyr, see Box 5.9)
Maximum treatments are one per year on non-crop land and two per year on established grasslands. The effects of treatment with triclopyr on Japanese knotweed are shown in Plate 17.

Imazapyr

(For a detailed account of imazapyr, see Box 5.10)
Imazapyr is a non-selective persistent herbicide and is therefore recommended for hard paving areas. Maximum treatments are one per year. This can be applied at any time during the growing season.

Picloram and triclopyr are selective for broad-leaved species and therefore allow a grass cover to persist throughout the time of treatment. However, due to their persistence in the soil, replanting with herbaceous and woody species is not possible for 3 months after using triclopyr and for 2 years after using picloram. Triclopyr and picloram are therefore particularly suited for use on paving, hard surfaces, non-amenity areas, cemeteries and derelict land with little wildlife value. Glyphosate and 2,4-D amine may also be used in these habitats (see Section 5.3.1). It is important to ensure that there is no regrowth in subsequent years following treatment. Further applications may be required.

In the UK picloram has been assigned an 'Occupational Exposure Standard'. This means that exposure by inhalation should not exceed 10 mg/m^3 over eight hours or 20 mg/m^3 over ten minutes.

BOX 5.8 Picloram factfile

NOT APPROVED FOR USE IN OR NEAR WATER

NB. When using herbicides, always read the product label and comply with the label instructions. It is an offence to use a herbicide in a manner which is not approved on the label. For requirements specific to your country or state, consult the agency responsible for pesticide regulation. Wear personal protective equipment as recommended by the product label.

Available as:	Soluble concentrate
Action:	A persistent translocated pyridine carboxylic acid herbicide for use on land not intended for cropping
Active ingredient:	240 g l^{-1} picloram (Tordon 22K)
Application rate:	For Japanese knotweed, 1008 – 1344 g active ingredient ha^{-1}
Method of application:	May be applied to active growth at any time of the year using a vehicle-mounted or knapsack sprayer
Time of application:	Best results achieved by application as foliar spray when the plant has a minimum height of 1 m Maximum number of treatments – 1 per year
Pre-treatment:	Cut back or trash any dead stems at the end of the season to facilitate access
Limitations:	Suitable for use where grass is to be retained; this is a selective herbicide which will have an effect on broad leaved species but not grasses This is a persistent herbicide and may restrict the growth of or kill broad leaved plants (including trees) planted or sown within 2 years of spraying On level ground there is negligible lateral movement, <blockquote>**but do not apply around desirable trees or shrubs where their roots may take up a lethal dose;**</blockquote>care should be taken on slopes to <blockquote>**prevent leaching into areas where desirable shrubs are present**</blockquote>
Effects:	Herbicide is translocated throughout the plant and causes the stems to twist and curl
Special requirements:	Use the product as specified on the label Avoid drift of spray on to desirable plants Prevent leaching into areas where desirable plants are present Do not spray if rain is falling or imminent Wear personal protective equipment as recommended by the product label Persistent in the soil for up to 2 years
Adjuvants:	None approved
Country specific information:	
UK:	Picloram is approved for use on land not intended for cropping; not approved for use in or near water
Denmark:	Not approved for use in Denmark according to the Danish Environment Protection Agency
Germany:	Not approved
USA:	Approved
New Zealand:	Approved

BOX 5.9 Triclopyr factfile

NOT APPROVED FOR USE IN OR NEAR WATER

NB. When using herbicides, always read the product label and comply with the label instructions. It is an offence to use a herbicide in a manner which is not approved on the label. For requirements specific to your country or state, consult the agency responsible for pesticide regulation. Wear personal protective equipment as recommended by the product label.

Available as:	Emulsifiable concentrate
Action:	A persistent translocated aryloxyalkanoic acid herbicide for use in forestry, non-crop areas and rough grazing
Active ingredient:	480 g l^{-1} (Garlon 4)
Application rate:	2880 g active ingredient ha^{-1}
Method of application:	Apply as an overall foliar spray during period of active growth when foliage is well developed using a vehicle-mounted or knapsack sprayer
Time of application:	Maximum number of applications – 2 per year
Pre-treatment:	Cut back or trash any dead stems at the end of the season to facilitate access
Limitations:	Suitable for use where grasses are to be retained; this is a selective herbicide which will have an effect on broad-leaved species but not grasses Best applied as a foliar spray A minimum interval of 6 weeks is required between application and planting
Effects:	Leaf browning and twisting and curling of stems (see Plate 17).
Special requirements:	Use the product as specified on the label Avoid drift of spray on to desirable plants Control may be reduced if rain falls within 2 hours of application Wear personal protective equipment as recommended by the product label Persistent in the soil for up to 3 months
Adjuvants:	Mixture B

Country specific information:

UK:	Triclopyr is approved for use in established grassland, rough grazing, non-crop areas, non-crop grass, roadside grass, forestry not for use in or near water. While Japanese knotweed is not specifically mentioned on the label, it is nevertheless approved for use as a foliar application for scrub clearance in forestry and industrial areas
Denmark:	Not approved for use in Denmark according to the Danish Environment Protection Agency
USA:	Approved
New Zealand:	Approved

BOX 5.10 Imazapyr factfile

NOT APPROVED FOR USE IN OR NEAR WATER

NB. When using herbicides, always read the product label and comply with the label instructions. It is an offence to use a herbicide in a manner which is not approved on the label. For require- ments specific to your country or state, consult the agency responsible for pesticide regulation. Wear personal protective equipment as recommended by the product label.

Available as:	Soluble concentrate
Action:	A non-selective translocated and residual imidazolinone herbicide for use in non-crop areas and forestry site preparation
Active ingredient:	imazapyr 50 gl^{-1} (Arsenal 50)
Application rate:	375 – 750 g active ingredient ha $^{-1}$
Method of application:	**AS A FOLIAR SPRAY** Knapsack sprayer Vehicle-mounted sprayer Controlled droplet applicator equipment
Time of application:	Apply at any time during the growing season while the plants are actively growing. Maximum number of treatments 1 per year
Pre-treatment:	Cut back or trash any dead stems at the end of the season to facilitate access
Limitations:	Imazapyr is a non-selective herbicide and is active on most plant species; it will provide long-term residual control of seedlings germinating after application
Effects:	Effects may not be visible for two weeks; complete kill of plants may not occur for several weeks
Special requirements:	Use the product as specified on the label Avoid drift on to desirable plants Do not use near desirable trees or shrubs or in areas where their roots may extend, or in locations where chemical may be washed or moved into contact with their roots. Do not apply to soil which may be used later to grow desirable plants Care must be taken when treatment is made on slopes where heavy rain after application may cause surface runoff following application Wear personal protective equipment as recommended by the label
Adjuvants:	Mixture B
Country specific information:	
UK:	Imazapyr is approved for use in non-crop areas where total vegetation control is to be achieved; it is also approved as a site preparation treatment in forestry Not approved for use in or near water
Denmark:	Not approved for use in Denmark according to the Danish Environmental Protection Agency
USA:	Imazapyr, e.g. Arsenal 50, can be applied at 4 – 6 pints/acre for Japanese knotweed control (the higher rate should be used where heavy or well-established infestations occur)

5.4 MECHANICAL CONTROL

A summary of non-chemical control methods is shown in Box 5.11.

5.4.1 Cutting

When carrying out cutting with a brush cutter or strimmer, protective clothing must be worn, including a face visor. It is essential to ensure that cutting
does not result in spreading Japanese knotweed from one site to another.
Machinery should be cleaned before leaving a site where Japanese knotweed
has been cut and any cut material should be collected either to dry out on

---

BOX 5.11 Summary of non-chemical control methods

Desired effect	Method	Timing	Frequency
Removal of dead stems Preparation of site prior to chemical treatment	Cutting with a metal bladed strimmer or trashing	Autumn/winter	Annually
Preventing invasion of an adjacent stand of Japanese knotweed into amenity grassland	Mowing or grazing	Throughout the growing season (March – October)	Mow every 2 weeks Allow livestock to graze throughout the growing season – prior to stocking
Reducing the vigour of the plant	Cutting or mowing	March – October	Four times a year
Removing individual stems of Japanese knotweed in mixed vegetation	Pulling	All year	As new shoots appear
Reducing plant height prior to chemical treatment	Cutting	March – August Allow regrowth to attain a height of 0.5–1.0 m (2–3 ft) before application of herbicide	As required

BOX 5.12 Cutting: a case study

Aims:
To establish the number of cuts required for the effective control of Japanese knotweed and to determine the effects of timing of cutting.

Location:
Greenhouse trials, USA.

Programme of action:
A number of pots were planted up with rhizome material of approximately 1 cm diameter and three nodes in length. The rhizomes were grown outdoors under ambient light and temperature condition The pots were selected at random for cutting at various intervals following establishment. Forty plants were selected for cutting at 28-day intervals between 5 June and 25 September. As a control 40 plants were allowed to grow on and were harvested on 15 October, 20 days after the last cut. At harvest, above and below ground material was separated, washed, oven-dried at 40oC and weighed. A further study looked at the effect of the number of cutting treatments. Pots were planted up with rhizome as described above. Thirty plants were selected randomly to each of eight groups: control; one cut (June, July or August) ; two cuts (June and July, June and August, or July and August) and three cuts (June, July and August).

Results:
Comparison of biomass of below ground material in cut plants showed that this was significantly lower than that of uncut controls (see table). The number of cuts made a significant difference to below-ground biomass, i.e. the more cuts the better, but the timing of cutting was not significant.

Results of cutting trials

Treatment	Below Ground Biomass
Control	31.24 g
Cut once	20.01 g
Cut twice	9.76 g
Cut three times	4.02 g

Lessons learned:
The timing of cutting does not influence the amount of below ground biomass at the end of the season.
The amount of rhizome biomass accumulated over the season was significantly less with each additional cut.
It is recommended that cutting should be carried out at least every four weeks during the growing season to prevent replenishment of below ground reserves in the rhizome. The timing of cutting do not appear to be critical so long as it is carried out at least seven weeks prior to senescence.
Cutting alone will not eliminate Japanese knotweed since the plant will continue to regenerate onc cutting is stopped. Cutting may be a useful treatment to reduce the vigour of Japanese knotweed ir combination with other treatments such as herbicide control.

References:
Seiger, L.A. and Merchant, H.C. (1997) Mechanical control of Japanese knotweed (*Fallopia japonica*): Effects of cutting regime on rhizomatous reserves. *Natural Areas Journal*, **17**, 341-345.

Seiger, L.A. (1993) The ecology and control of *Reynoutria japonica (Polygonum cuspidatum)*. Ph.D. Thesis, The George Washington University, Washington, DC.

Scott, R. and Marrs, R.H. (1984) Impact of Japanese knotweed and methods of control. *Aspects of Applied Biology* **5**, 291-296.

site or to be removed for disposal at licensed landfill according to current guidelines (Chapter 6). It is essential that cut material is prevented from entering watercourses. A case study of cutting Japanese knotweed is presented in Box 5.12.

- Frequent cutting may reduce the vigour of Japanese knotweed after several years provided the plant is not invading from elsewhere.
- Use a lopper or brush cutter with a metal circular blade.
- The first cut can be undertaken at the burst of spring growth (April - May).
- The last cut can be made in late summer when the plant is most luxuriant and before it starts to die back (September).
- This cutting regime should be repeated annually to maintain a level of control.

Care should be taken to monitor the site to ensure that cutting has not encouraged the plant to spread sideways, as there is some evidence of increased lateral rhizome growth and increased stem density after cutting.

WARNING

Cut material has the potential to regenerate into new plants and should therefore be prevented from entering watercourses including ditches. Cut material should be either piled on site and inspected regularly for signs of re-growth or removed from site and disposed of according to any regulations with respect to Duty of Care (see Section 2.6 and Chapter 6).

5.4.2 Mowing

In amenity areas, verges, parks and gardens, mowing with a conventional grass-mower can be a reasonable method of control whilst the mowing regime is in force. Frequent mowing at fortnightly intervals during the growing season may also prevent invasion of grassed areas. If regular mowing is stopped Japanese knotweed may take over a site again as it is not yet known how persistent the rhizomes are in the soil. Flail-mowing is not recommended as it may result in fragments of viable stem material being spread into non-infested areas.

5.5 MANUAL CONTROL

For small stands and in areas of high ecological value, where the use of herbicides is thought to be unsuitable or is prohibited, manual methods of control are recommended.

5.5.1 Pulling

Pulling up mature stems complete with roots over a 3-year period has been shown to eliminate small infestations of Japanese knotweed, giving other vegetation a competitive advantage. All visible stems must be uprooted by pulling from the base of the stem when at their full height. A case study of pulling Japanese knotweed is presented in Box 5.13.

BOX 5.13 Pulling: a case study

Aims:
To eradicate a small infestation of Japanese knotweed in sensitive areas.

Location:
A local nature reserve, Mersey Valley, Warrington, United Kingdom.

Programme of action:
A small infestation of Japanese knotweed measuring approximately 2 square metres was identified at the trial site. Individual stems were hand-pulled removing as much of the rhizome as possible with each stem when plants were at full height in August. Pulling was continued for a number of years until no new shoots were recorded. Care was taken to ensure proper disposal of the pulled stem material to avoid further spread. The stems were piled and allowed to dry out completely and were then burned on site.

Results:
A complete control of the plant was achieved after continued pulling throughout the growing season over a 3-year period.

Lessons learned:
- Pulling stems from the base can be an effective method of control only if carried out continually for a number of years.

- This method is only applicable to small infestations and is ideally suited to new infestations when only a few stems have established.

- The use of volunteers who are willing continually to return to the site to carry out the treatment an advantage.

- As this method targets only the Japanese knotweed plants, it is suitable for the protection of native/sensitive species growing in the vicinity, or on sensitive sites where the use of herbicides undesirable.

- Trampling damage may occur to other plants in the immediate vicinity whilst the treatment is being carried out.

References:
Baker, R.M. (1988) Mechanical control of Japanese Knotweed in an S.S.S.I. *Aspects of Applied Biology*, **16**, 189-192.

WARNING

Stem and rhizome material has the potential to regenerate into new plants and should therefore be prevented from entering watercourses including ditches or being taken off site inadvertently (i.e. with other plant waste). Japanese knotweed material should be piled either on site and inspected regularly for signs of regrowth or removed from site and disposed of according to any regulations with respect to Duty of Care (see Section 2.6 and Chapter 6).

5.5.2 Ineffective control measures

- Burning the actively growing plants in situ has not been found to be an effective method of control and is not recommended.

- Deep digging well-established areas of Japanese knotweed results in a significant increase in stem density. However, when combined with herbicide treatment this can be an effective control measure (see Section 5.7.3). Fragmentation of the rhizome system enhances the production of new shoots. Equipment which has been used to dig areas of Japanese knotweed is likely to become contaminated with small plant fragments which may infect new sites. It is essential that any digging activity in the vicinity of Japanese knotweed plants is followed by a careful examination and cleaning of all equipment prior to leaving the site.

5.6 BIOLOGICAL CONTROL

Grazing by livestock is a viable option for controlling Japanese knotweed and has been used in Europe and North America. There is also the potential for control of the species using classical biological control.

5.6.1 Grazing

Japanese knotweed is palatable to sheep, cattle, horses, donkeys and goats and was originally introduced to parts of Europe as a plant suitable for fodder. Grazing occurs mostly in the spring as young shoots are preferred. The availability of young shoots declines after late July and the plants become rather woody. Grazing may reduce shoot densities and shoot height but will not eradicate Japanese knotweed although it may reduce the spread of the plant into uninfested areas. Control will only be achieved if animals are allowed to graze the Japanese knotweed for the whole of the growing season, commencing early in the year before the plants become tall and woody. The presence of dead stems is a deterrent to grazing, it is necessary to cut and remove the previous year's growth before grazing commences. Cutting several times during the season will ensure that young shoots continue to be available. Disposal of cut material must be carried out safely (Section 6).

Intensive grazing in Japanese knotweed-free areas may prevent infestation from adjacent areas but this method will only suppress the growth of Japanese knotweed and will not eradicate the plant.

BOX 5.14 Comparison of insect herbivore taxa on Japanese knotweed in the UK and Japan

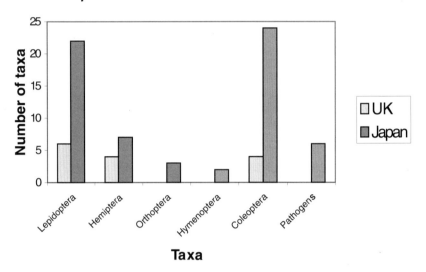

Data from: Greaves, M. and Shaw, R. (1997) *Biological control of weeds, a scoping study of the feasibility of biological control of aquatic and riparian weeds in the UK*. Environment Agency R&D Tech. Report W 105.

WARNING

Animals must not be allowed to graze on Japanese knotweed that has been treated with herbicide until after the time period recommended by the manufacturer on the label.

5.6.2 Bio-control agents

Clonal plants, such as Japanese knotweed, have in the past made good candidates for biological control, and there is potential for future control by biological agents. In its native range, Japanese knotweed is controlled by herbivores and/or disease and is not a problem weed. As with many introduced species which arrive in their new range without associated pests and pathogens, Japanese knotweed has few or no pathogens and predators associated with it in Europe and North America.

A bio-control programme relies on a reduction in the presence of the plant to a more acceptable level rather than eradication. The advantages of such a programme would be the permanence and sustainability of control and the absence of harmful side-effects associated with herbicide control especially in such sites as water protection areas, nature reserves and recreational areas. The development of a bio-control programme for Japanese knotweed will take a number of years.

A classical biological control programme would begin with the selection of control organisms from the natural enemies of the plant in Japan, Taiwan or northern China. Once potential control agents have been identified, extensive testing for host specificity would be undertaken to avoid the organism adopting other plant species as hosts. Culturing and quarantine procedures would be carried out to build up numbers and to free the control organism from parasites and disease.

Following initial comparisons of invertebrates recorded on the plants in Japan and the UK (Box 5.14), there would appear to be scope for biological control of Japanese knotweed. This could be achieved by using Japanese invertebrate species such as beetles. An alternative method would be to use an invertebrate species indigenous to areas in which it has become naturalized, which could be bred specifically with a preference for Japanese knotweed.

Several pathogens which cause significant damage to Japanese knotweed in Japan have been identified. These include rust fungi, two species of *Puccinia* and a leaf spot fungus, *Phyllosticta rayoutina.*

5.7 INTEGRATING CONTROL OPTIONS

Control methods which integrate a variety of options can lead to effective control of Japanese knotweed.

5.7.1 Cutting and spraying

Cutting the plant in late spring or early summer and waiting for regrowth prior to treating with herbicide may reduce the vigour of the plant and provide a more manageable height at which to spray. A case study describing a combination of cutting and spraying is presented in Box 5.15.

5.7.2 Combination of chemical treatments

Spraying with a non-selective herbicide initially such as glyphosate followed by spraying with a selective herbicide such as 2,4-D amine will aid the recovery of a replacement grass sward.

5.7.3 Combination of digging and spraying

A combination of treatments may provide a more effective level of control than using each individually. A case study describing a combination of digging and spraying is presented in Box 5.16.

BOX 5.15 Combination treatments – cutting and spraying: a case study:

Aims:
To eradicate a large infestation of Japanese knotweed adjacent to a river.

Location:
Floodplain of the River Tawe near Pontardawe, Wales, UK.

Programme of action:
An extensive infestation of Japanese knotweed in the floodplain of the River Tawe was identified as the trial site. Treatment plots measuring approximately 100 square metres were set up as follows:

Plot	Treatments		
Control	no treatment		
Plot 1	no cut	+	sprayed twice using knapsack sprayer
Plot 2	cut once	+	sprayed once using knapsack sprayer
Plot 3	cut once	+	no spray
Plot 4	no cut	+	spray using telescopic lance sprayer

As the plots were near water, the herbicide chosen for the trial was glyphosate which was applied at a rate of 5 l/ha (2.4 kg active ingredient per hectare). Cutting was carried out with a brush cutter with a metal blade at the begining of the trial (March) to remove dead stems and in July for the cutting treatment. Plots were monitored for regrowth over a two-year period by visual assessment of plant cover.

Results:

Plot	Treatment			% Cover after 2 Years
Control	no treatment			100
Plot 1	no cut	+	sprayed twice using knapsack sprayer	<1
Plot 2	cut once	+	sprayed once using knapsack sprayer	5 – 10
Plot 3	cut once	+	no spray	30 – 40
Plot 4	no cut	+	spray using telescopic lance sprayer	5 – 10

- Treatment with glyphosate applied via a knapsack sprayer (sprayed twice) reduced cover from 100% to <1% over the two-year period.

- A single cut followed by a single spray of glyphosate reduced cover from 100% to 5 – 10% in two years.

- Treatment with glyphosate applied via the telescopic lance sprayer achieved good control over the two-year period, reducing the total cover of Japanese knotweed to 5 – 10% compared with the control plot.

- Cutting only reduced the percentage cover to 30 ñ 40% compared to 100% cover for the control.

Lessons learned:
For stands of Japanese knotweed with difficult access, the use of a telescopic lance sprayer is an effective means of herbicide application.

The effect of cutting reduces total cover but is not an effective treatment alone. When combined with a herbicide application the results are improved. However it is necessary to apply glyphosate more than once per year to achieve effective control.

It is necessary to continue treatment for more than two years in order to achieve complete control.

References:
de Waal, L.C. (1995) Treatment of *Fallopia japonica* near water. In: Pysek, P., Prach, K., Rejmanek, M. and Wade, P.M., (eds), *Plant Invasions General Aspects and Special Problems*. SPB Academic Publishing, Amsterdam. pp. 203-212.

BOX 5.16 Combination treatments – digging and spraying: a case study

Aims:
To develop a rapid treatment for the control of large areas of Japanese knotweed.

Location:
An area of waste industrial land adjacent to the Channelsea River at Mill Meads, Stratford, east London, UK.

Programme of action:
An area of monoculture Japanese knotweed on waste industrial ground adjacent to the Channelsea River in east London was identified as the trial site. The area was divided into sub-plots for a range of treatments including deep digging by mechanical excavator, foliar spray with the herbicide glyphosate via a knapsack sprayer and a combination of the two treatments. A control plot was allocated to receive no treatment. During the three-year trial, the plots were monitored at 7, 12 and 30 months and measurements of stem density, shoot height and stem diameter were taken.

Pre-treatment:
The site was cleared in late September using a metal-bladed strimmer to clear old stems. Dead stem material was allowed to remain in situ.

Digging:
Plots allocated to the digging treatment were excavated in October to a depth of 0.5 m (18 in) using a mechanical excavator. The dug material was turned and respread over the plot to create a level surface.

Spraying:
Plots allocated to the spraying treatment were treated with glyphosate at a rate of 1800 g ha^{-1} active ingredient in May and in July the following year.

Results:
Results are shown in the table below. The combination of digging and spraying resulted in a more rapid and significantly more effective control than spraying alone over the 3-year period of the trial. Digging alone resulted in a significant increase in stem density although plants in these plots showed a smaller stem diameter and were shorter in height than in control plots. A single application of glyphosate had little effect on plants subjected to the spraying treatment only. Two applications of glyphosate showed a 79% reduction in stem density over a 30-month period on sprayed only plots but the greatest level of control was achieved by combined digging and spraying treatments. Digging followed by a single spray gave 93% reduction in stem density within 19 months whereas digging followed by two spray treatments resulted in 98% reduction in stem density by the end of the trial. The combination of digging and spraying also allowed the natural establishment of other species which were suppressed by monoculture Japanese knotweed on other plots.

Lessons learned:
- Deep digging encourages shoot production and increases stem density through fragmentation of the rhizome system.

- Treatment with herbicide alone, requires application over a number of years to achieve effective control.

- A combination of deep digging followed by herbicide application effectively increases the above: below ground biomass ratio of the plant and allows a more even delivery of herbicide via translocation to fragmented rhizomes.

- Combined digging and spraying treatments reduce the time required to achieve an effective level of control.

- Relative treatment costs are shown in Box 2.21.

Treatment	No digging	Digging
No spraying	0	Increase by 200%
Sprayed once with glyphosate	25% reduction	93% reduction
Sprayed twice with glyphosate	79% reduction	98% reduction

Effect of combination treatments shown as percentage reduction of stem density in treated plots compared to control plots.

Reference: Child, L.E., Wade, P.M. and Wagner, M. (1998), see Bibliography p.97.

5.8 SITE REVEGETATION

An important aspect of control of Japanese knotweed is to anticipate the problems associated with removal of large and/or dense stands before they occur. As Japanese knotweed outshades other vegetation it is possible that with a non-selective herbicide treatment, the rapid removal of vegetation will lead to one of the main problems, namely soil erosion. It is therefore necessary to build into the treatment programme a strategy for revegetation. This could entail changing the herbicide treatment during the programme from glyphosate initially to a selective herbicide such as 2,4-D amine for subsequent treatments to allow a grass cover to persist. Alternatively, plants remaining after an initial treatment with glyphosate could be spot-treated using a weed-wiper or similar application to allow other plants to establish successfully. Whatever the strategy adopted, whether planting with selected species or allowing vegetation to establish naturally, it should be included as an important element of the control strategy and should be determined at the start of the programme. Species suitable for revegetation include use of grass-seed mixes. Planting of shrubs should be carefully considered as monitoring the site for regrowth of Japanese knotweed following treatment could be impeded. Re-application of herbicide to Japanese knotweed, if necessary, may adversely affect trees and shrubs and reduce treatment options in future.

Summary

- There are a number of different treatment options for the control of Japanese knotweed, the choice of which is dependent on:
  - situation/ location;
  - size of infestation;
  - site access;
  - presence of other vegetation;
  - cost.
- It is essential to assess the impact of any treatment options prior to implementation.
- Site revegetation following treatment is an important component of a successful management strategy.
- The control of Japanese knotweed needs to take into consideration an integrated approach to the problem.
- Over the coming years it is hoped that biological control will provide a sustainable method of control.
- Extreme care is needed to ensure that control methods do not result in further spread of Japanese knotweed to new sites.

6   Disposal options

'Let's talk of graves...' (William Shakespeare)

The options available for disposal of Japanese knotweed waste, for example, stems, leaves and rhizomes are:

- On-site burning,
- On-site burial,
- Off-site burial.

6.1 DISPOSAL OF SOIL CONTAMINATED WITH RHIZOMES

Current regulations for the disposal of soil containing rhizome fragments must be adhered to. Contact your local environment agency waste regulations department to find out the current procedures for transportation of spoil and licensed disposal sites.

If possible, disposal on site is the best option. This reduces the risks of spreading the plant during transportation, reduces the costs associated with landfill charges and reduces the need for importation of spoil to sites where development is due to take place. Disposal on site also ensures that the effectiveness of control can be monitored over the treatment period ensuring complete eradication.

6.2 DISPOSAL OF STEM MATERIAL

Cut plant material should preferably be burned on site, in a manner which complies with local or national legislation on burning. In the UK, for example, the Ministry of Agriculture Fisheries and Food provides guidelines for farmers when burning crop residues. These point out that if burning is carried out carelessly prosecution can follow under a variety of laws. If it is decided to dispose of cut plant material in a landfill that must comply with guidelines too.

Cut plant material could be placed in small piles on site to dry out and eventually die. The site should be inspected regularly to ensure at all times that the stems cannot be blown or washed into a watercourse or produce new plants due to wet conditions. It is not recommended that cut stems of Japanese knotweed are added to compost.

6.3 CURRENT GUIDELINES AND LEGISLATION

When dealing with any aspect of Japanese knotweed, it is essential that due regard is taken of current guidelines and legislation. For example, the current legislation relating to the plant in the UK is explained in Section 2.6. To date, no legislation relates to this particular plant in other European countries. However, individual state laws on noxious plants in the USA and the Biodiversity Convention in the European Union may change this position. It is recommended that you find out from your local environment agency the current situation regarding the plant, its control and disposal, and make sure that you abide by current legislation. The transportation and spread of this plant has been aided greatly in the past by human activity. Make sure that you are not responsible for its further spread.

There is a need to develop a reliable method for the decontamination or removal of vegetative fragments of Japanese knotweed from soil or spoil. An economic viability study for the Welsh Development Agency looked at a range of options (see Box 6.1).

BOX 6.1 Options for treatment of soil contaminated with Japanese knotweed rhizome fragments

- **Soil sterilization** – use of chemicals such as methyl bromide

- **Soil freezing** – use of liquid nitrogen

- **Soil burning** – 'on-site' incineration

- **Soil sieving** – screening soil to remove fragments of rhizome material

---

Summary

- **Follow guidelines and laws relating to the disposal of Japanese knotweed material.**

- **The main options are burning or disposal to landfill.**

- **Prevention is better than cure – make sure that the plant is not spread by vegetative means by adopting a policy for proper disposal and treatment of material from infested sites.**

7 Glossary

Terms printed in **bold type** within an explanation have their own explanations. Please cross-refer.

achene	Seed capsule
active ingredient	The concentration of an herbicidally-active chemical within a formulation.
adjuvant	A substance or substances which, when added to a pesticide, increases the efficacy of the treatment.
adventitious	Concerning roots which develop from part of the plant other than the roots, e.g. in Japanese knotweed from the stem.
alien	A plant whose native range lies outside the home country, e.g. the **British Isles**. Also see **naturalized**.
alluvial	Of or relating to a fine-grained fertile soil (allivium) consisting of mud, silt and sand deposited by flowing water.
annual	A plant that completes its life-cycle from germination as a seed to production of seed followed by death within a single year.
anther	The terminal portion of a **stamen** containing pollen.
autecology	The study of the relationships between a single species and its environment.
axil	The angle between the leaf stalk of a plant and the stem.
axillary	The term used to describe that part of a plant occuring in the **axil** of a leaf
bio-control	See **biological control**.
biological control	Artificial control of pests and diseases by using other organisms, e.g., introducing a fungal disease of Japanese knotweed from Japan to Europe in order to control the plant there.
biomass	The quantity or weight of living material.
British Isles	The geographical area including the **United Kingdom**, the Isle of Man and the Republic of Ireland.

carpel	The female reproductive parts of flowering plants consisting of ovary containing ovules which become seeds after fertilization, and **stigma** and **style**.
catchment management plan	A plan developed for the management of the land, river or stream/creek within the catchment of a watercourse.
chromosome	A rod-shaped portion of that part of the nucleus of a cell which is fundamental to the division of cells and the transmission of hereditary characteristics from one generation to the next.
clone	Genetically identical individuals.
cortex	The outer layer of tissue in roots and stems bounded on the outside by the epidermis.
cross	The act or product of cross-fertilization.
crown	The visible part of the rhizome from which a bud or buds will erupt.
database	A body of information stored, usually in a computer, from which particular items of information can be retrieved when required.
dioecious	Applied to organisms in which male and female reproductive organs are borne by different individuals.
dormancy	A resting condition in which the growth of an organism is halted and **metabolic rate** is slowed down.
Duty of Care	Regulations requiring that waste which may cause harm to the environment (including waste which could cause interference with ecological systems) is disposed of safely according to appropriate legislation, e.g. the United Kingdom *Environmental Protection Act* (1990). Transportation, handling and final disposal of such waste must be carried out by licensed operators, and written records must be kept.
ecosystem	A unit consisting of a community of organisms and the environment in which they live and also interact, e.g. a wood.
family	A group of similar **genera** of taxonomic rank below order and above **genus**; with plants, the names usually end in –aceae, e.g. Polygonaceae.
flexuous	Bending.
fumarole	A hole in the ground emitting gases in a volcano or volcanic region.

garden	A piece of ground adjoining a house where flowers and/or vegetables are cultivated.
genus	A taxonomic group lower than a **family** consisting of closely related species, e.g. *Fallopia*. The convention is to use the initial letter as an abbreviation, e.g. *F. japonica*.
genera	Plural of **genus**.
Geographical Information System	(abbreviated to GIS) A powerful set of tools for collecting, storing, retrieving at will, transforming and displaying data from the real world geophyte.
herbaceous	Usually refers to tall herbs that die down during the winter and survive as underground or **perennating** parts till conditions improve.
herbicide	A chemical used to kill weeds.
herbivore	Plant-eating organism.
hybrid	A plant originating after the fertilization of one species by another species.
inflorescence	A flowering shoot or an aggregation of flowers.
internode	Between **nodes** – knob on root or branch from which leaves arise.
introduced	In this context, refers to the process of either knowingly or unknowingly bringing a plant into an area outside its native range.
invertebrate	A collective name for all those animals which do not possess backbones or vertebrae such as insects and spiders.
journal	A publication, usually published quarterly, of papers or articles of a learned nature.
lignin	A constituent of cell walls in plants which imparts rigidity.
local authority	Self-administration in local affairs by towns, cities and communes as distinct from national, federal or central government.
logarithmic scale	A scale of measurement in which an increase or decrease of one unit represents a ten-fold increase or decrease in the quantity measured, e.g. **pH**.
lowland	Relatively low ground or a low generally flat region.

metabolic rate	The rate at which the totality of chemical changes occur in an organism such as a plant.
midrib	The main vein or nerve of a leaf.
monoculture	The cultivation of a single crop.
monophagous	Applied to an animal which feeds only on one type of food.
naturalized	An introduced species that is permanently established (naturalization).
native	Belonging naturally to a region, e.g. Japanese knotweed is native to Japan.
nectary	Gland secreting sugary fluid, that is nectar.
node	A place, often swollen where one or more leaves are attached to a stem.
nutrient	A nourishing substance, in this context with respect to plant growth.
ornamental	A plant grown for ornament or beauty.
panicle	A branched **inflorescence**.
parasite	Organism living in or on another organism (its host) from which it obtains food.
pathogen	An organism or substance that causes disease.
perennation	Survival from year to year by vegetative means. The vegetative parts of the plant which survive the inclement period are known as the **perrenating** parts, e.g. the rhizome of Japanese knotweed.
perennial	A plant that continues its growth from year to year.
petiole	A leaf-stalk.
pH	A quantitative expression for acidity or alkalinity of a solution. The **logarithmic scale** ranges from 0 (acid) to 14 (alkaline).
phenology	Study of periodicity phenomena of plants, e.g time of flowering in relation to climate.
photosynthesis	The use of energy from sunlight to drive chemical reactions in green plants that lead to the formation of sugars, e.g. glucose, and starch.
phytophagous	Plant-eating.

pollinator	An organism which effects the transference of pollen from **anther** to **stigma**, usually an insect.
polyphagous	Feeding on many types of food.
predation	Relationship between two species of animal in a community in which one hunts, kills and eats the other.
press release	An official report or statement supplied to the press.
propagule	Any part of a plant capable of growing into a new organism, e.g. spore, seed or cutting.
rhizomatous	Of or pertaining to a **rhizome**.
rhizome	An underground stem, bearing buds and scale leaves, which lasts for more than one season and usually serves for both vegetative propagation and **perennation**.
riparian	Of or inhabiting a riverbank.
ruderal	Applied to plants which inhabit old fields, waysides or waste land.
section	A taxonomic group which is the sub-division of a **genus**, e.g. *Parogonum* is a section of the genus *Fallopia* (see Box 2.6).
semi-natural	That which is partly modified by human influence.
Site of Special Scientific Interest	(abbreviated to SSSI). Designation for an area of land in the United Kingdom which has been notified by English Nature, the Countryside Council for Wales or Scottish Natural Heritage as being of special interest for its flora, fauna or geological or physiogeographical features in order to confer protection on the site.
species	A taxonomic group of individuals having common characteristics and placed under a **genus**.
stamen	The pollen-producing part of a flower consisting of the **anther** and the filament.
stand	A group of individual plants usually of the same **species** growing in close proximity.
stigma	That part of the plant, the surface of the **carpel**, which receives the pollen.
style	A stalk-like structure of the **carpel** bearing the **stigma**.
sulphur dioxide	A suffocating gas discharged into the atmosphere as a part of volcanic processes and in waste from industrial processes.
taxa	See **taxon**.

taxon	A biological category, a taxonomic group or unit, e.g. **family**, **genus**, **species** and **variety**.
taxonomy	The classification of plants and animals.
translocate	The process by which materials, e.g. **nutrients**, are transported within a plant.
United Kingdom	Since 1922, the official title for the kingdom consisting of England and Wales, Scotland and Northern Ireland (abbreviated to UK).
understorey	Collectively the trees in a forest below the upper canopy.
variety	The **taxon** below species; a group which distinctly differs, for various reasons, from other varieties within the same species, e.g. *Fallopia japonica* var. *japonica* is a variety of the species *Fallopia japonica*.
vermiculite	A water-absorbent substance used for planting seeds and plants; based on a form of mica.
viability	Being capable of germinating, living and surviving.
woody	Of tissues which are hard because of **lignin** in the cell walls.
yard	An enclosed space or **garden** adjacent to a building where flowers and/or vegetables are grown.

8   Bibliography

8.1 ADVANTAGES OF JAPANESE KNOTWEED

Anantaphruti, M., Terada, M., Ishii, A.I., Kino, H., Sano, M., Kuroyanagi, M. and Fukushima, S. (1982) Studies on chemotherapy of parasitic helminths (XI). *In vitro* effects of various drugs on the motor activity of adult *Schistosoma japonicum*. *Japanese Journal of Parasitology*, **31**, 321-328.

Arichi, H., Kimura, Y., Okuda, H., Baba, K., Kozawa, M. and Arichi, S. (1982) Effects of stilbene components of the roots of *Polygonum cuspidatum* Sieb. and Zucc. on lipid metabolism. *Chem. Pharm. Bull.*, **30** (5), 1766-1770.

Callaghan, T. V., Scott, R., Lawson, G.J. and Mainwaring, A.M. (1984) An experimental assessment of nature and naturalised species of plants as renewable sources of energy in Great Britain: 3; Japanese knotweed (*Reynoutria japonica*). *Bulletin of the Energy Technology Support Unit*, ETSU-B 1086/3.

Callaghan, T. V., Scott, R., Lawson, G.J. and Proctor, A.M. (1984) Experimental assessment of native and naturalised species of plants as renewable sources of energy in Great Britain. *E.I. Monthly*, EIM8611-07244.

Callaghan, T. V., Scott, R., Lawson, G.J. and Mainwaring, A.M. (1984) An experimental assessment of native and naturalised species of plants as renewable sources of energy in Great Britain. Japanese knotweed – *Reynoutria japonica*. Institute of Terrestrial Ecology Project 674. National Environment Research Council, London.

Callaghan, T. V., Scott, R. and Whittaker, H.A. (1981) The yield, development and chemical composition of some fast-growing indigenous and naturalised British plant species in relation to management as energy crops. Institute of Terrestrial Ecology Project 640, National Environment Research Council, London.

Chang, C.J., Ashendel, C.L., Geahlen, R.L., McLaughlin, J.L. and Waiters, D.J. (1996) Oncogene signal transduction inhibitors from medicinal plants. *Vivo* (Attiki), **10** (2), 185-190.

Daayf, F., Schmitt, A. and Belanger, R.R. (1995) The effects of plant extracts of *Reynoutria sachalinensis* on powdery mildew development and leaf physiology of long English cucumber. *Plant Diseases*, **79** (6), 577-580.

Ezaki, T. and Fushima, T. (1976) Studies on the use of Japanese weeds for the protection work of bare slope. (II) Examination on the cutting slope of the forest road. *Bulletin of the Ehime University Forest*, **13**, 161-174.

Farrazzi, P. and Marletto, F. (1990) Bee value of *Reynoutria japonica* Houtt. *Apicoltore Moderno*, **81** (2), 71-76.

Hiraoka, A. and Yoshimata, K. (1986) Isotachophoresis of anthocyanins. *Chem. Phar. Bull.*, **34** (5), 2257-2260.

Horigome, R., Kumar, R. and Okamoto, K. (1988) Effects of condensed tannins prepared from leaves of fodder plants on digestive enzymes in vitro and in the intestines of rats. *British Journal of Nutrition*, **60** (2), 275-285.

Howes, F.N. (1979) *Plants and beekeeping*. Faber & Faber, London, 184pp.

Hyun, I. W., Lim, K.H., Shin, M.S., Won, YJ., Kim, Y .S., Kang, S.S., Chang, I.M., Woo, W.S., Paik, W.H., Kim, H.J., Woo, E.R., Park, H.K. and Park, J.G. (1994) Antineoplastic effect of extracts from traditional medicinal plants and various plants. *Korean Journal of Pharmacognosy*, **25** (2), 171-177.

Jayasuriya, H., Koonchanok, N.M., Geahlen, R.L., McLaughlin, J.L. and Chang C.J. (1992) Emodin, a protein tyrosine kinase inhibitor from *Polygonum cuspidatum*. *Journal of Natural Products*, **55** (5), 696-698.

Jayatilake, G.S., Jayasuriya, H., Lee, E.-S., Koonchanok, N.M., Geahlen, R.L., Ashendel, C.L., McLaughlin, J.L. and Chang, C.J. (1993) Kinase inhibitors from *Polygonum cuspidatum* – Bioassay-directed fractionation of a medicinal plant. *Journal of Natural Products* (Lloydia), **56** (10), 1805-1810.

Kim, T .H. and Lee, C.H. (1973) Pharmacognostical studies on *Polygonum* spp. *Korean J. Pharmacogn.*, **4** (2), 75-82.

Kimura, Y., Kozawa, M., Babam K. and Hata, K. (1983) New constituents of roots of *Polygonum cuspidatum*. *Planta Medica*, **48** (3), 164-168.

Kimura, Y., Okuda, H. and Kubo, M. (1995) Effects of stilbenes isolated from medicinal plants on arachidonate metabolism and degranulation in human polymorphonuclear leukocytes. *Journal of Ethnopharmacology*, **45** (2), 131-139.

Kubo, M., Kimura, Y., Shin, H., Haneda, T., Tani, T. and Namba, K. (1981) Studies on the antifungal substance of crude drug (II). On the roots of *Polygonum cuspidatum* Sieb. et Zucc. (Polygonaceae). *Shoyakugaku Zasshi*, **35** (1), 58-61.

Kubota, K., Nishizono, H., Suzuki, S. and Ishii, F. (1988) A copper-binding protein in root cytoplasm of *Polygonum cuspidatum* growing in a metalliferous habitat. *Plant Cell Physiology*, **26**, 1029-1033.

Kurokawa, M., Nagasaka, K., Hirabayashi, T., Uyama, S.I., Sato, H., Kageyama, T., Kadato, S., Ohyama, H., Hozumi, T., Namba, T. and Shiraki, K. (1995) Efficacy of traditional herbal medicines in combination with acyclovir against herpes simplex virus type 1 infection *in vitro* and *in vivo*. *Antiviral Research*, **27** (1-2), 19-37.

Kuznetsova, Z.P. (1979) Study of phenolic compounds from Japanese knotweed. *Vesli Akademy Navuk Belarus SSR*, **5**, 29-32.

Leung, A.W.N. (1996) Protective effects of polydatin, an active compound from *Polygonum cuspidatum*, on cerebral ischemic damage in rats. *Chinese Pharmacological Bulletin*, **12** (2), 126-129.

Lin, Ming Hsuan; Hsu, Shih Yuan (1987) Studies on pharmacological effects of various extracts of *Polygonum cuspidatum* S et Z. *Taiwan Yao Hsueh Tsa Chi*, **39** (1), 42-53.

Masaki, H., Sakaki, S., Atsumi, T. and Sakurai, H. (1995) Active-oxygen scavenging activity of plant extracts. *Biological and Pharmaceutical Bulletin*, **18** (1), 162-166.

Matsuda, K. (1976) Flavonoids as feeding stimulants of the beetles attacking the Polygonaceous plants. *Tohoku Journal of Agricultural Research*, **27** (3-4), 115-121.

Matsumura, T., Hayakawa, H. and Hasegawa, T. (1981) Oviposition of *Chrysops suavis* Loew (Diptera, Tabanidae) with special reference to its preference for plant and site. *Bulletin of the National Grassland Research Institute*. **19**, 56-65.

Mende, A., Mosch, J. and Zeller, W. (1994) On the induced resistance of plant extracts of fire blight *(Erwinia amylovora)*. *Journal of Plant Diseases and Protection*, **101** (2), 141-147.

Mino, Y., Usami, H., Ota, N., Takeda, Y., Ichihara, T. and Fujita, T. (1990) Inorganic chemical approaches to pharmacognosy. VII. X-ray flurorescence spectrometric studies on the inorganic constituents of crude drugs. (5) The relationship between inorganic constituents of plants and the soils on which they are grown. *Chem. Pharm. Bull.*, **38** (8), 2204-2207.

Murakami, T. (1968) On the structures of the anthraglycoside from the rhizomes of *Polygonum cuspidatum* Sieb. et Zucc. *Chem. Pharm. Bull.*, **16** (11), 2299-2300.

Murakami, T. and Tanaka, K. (1973) Water soluble polysaccharide from the roots of *Polygonum cuspidatum* Sieb. et Zucc. *Chem. Pharm. Bull.*, **21** (7), 1506-1509.

Okomoto, K., Yamamoto, Y. and Fuwa, K. (1978) Accumulation of manganese, zinc, cobalt, nickel and cadmium by *Clethra barbinervis*. *Agricultural and Biological Chemistry*, **42** (3), 663-664.

Pellett, F.C. (1943) The Postscript. *American Bee Journal*, **83** (11), 438.

Plass, W.T. (1975) An evaluation of trees and shrubs for planting surface-mine spoils. *USDA Forest Service Research Paper*, NE-317. Northeastern Forest Experiment Station. pp. 8.

Pneva, G.P. (1989) On *Reynoutria* spp. from the Neogene of Primorski Krai Russian SFSR USSR and Japan. *Botanicheskii Zhurnal* (Leningrad), **74** (7), 1032-1033.

Schneider, S. and Ullrich, W.R. (1994) Differential induction of resistance and enhanced enzyme activities in cucumber and tobacco caused by treatment with various abiotic and biotic inducers. *Physiological and Molecular Plant Pathology*, **45**, 291-304.

Stevens, W.A. and Reynolds, T. (1992) Plant virus inhibitors from members of the Polygonacaea. *Biomedical Letters*, **47**, 269-273.

Su, Hsueh-Yueh, Shur-Hueih Cheng, Chien-chung Chen and Huei Lee (1995) Emodin inhibits the mutagenicity and DNA adducts induced by 1-nitropyrene. *Mutation Research*, **329** (2), 205-212.

Sugahara, T., Yamaguchi, H., Sasaki. H. and Aoyagi. Y. (1988) Mineral contents of newly commercialized vegetables and herbs. *Joshi Eiyo Daigaku Kiyo*, **19**, 131-138.

Tahara, S., Matsukura, Y., Hirohuki, K. and Mizutani, J. (1993) Naturally occurring antidotes against benzimidazole fungicides. *Zeitschrift für Naturforschung. Section C Biosciences*, **48** (9-10), 757- 765.

Wiltshire, E.P. (1970) An apparently unrecorded floral attraction for moths and other insects in September. *Entomologist's Record*, **82**, 301-2.

Yamaki, M., Asogawa, T., Kashihara, M., Ishiguro, K. and Takagi, S. (1988) Screening for antimicrobial action of Chinese crude drugs and active principles of Hu Zhang. *Shoyakugaku Zasshi*, **42** (2), 153-155.

Yeh, S.F., Chou, T.-C. and Liu, T.-S. (1988) Effects of anthraquinones of *Polygonum cuspidatum* on HL-60 cells. *Planta Medica*, **54**, 413-414.

Yoshimata, K., Nishino, H., Ozawa, H.. Sakatani, M., Okabe, Y. and Ishikura, N. (1987) Distribution pattern of anthocyanidins and anthocyanins in Polygonaceous plants. *Botanical Magazine Tokyo*, **100** (1058), 143-150.

8.2 BIOLOGICAL CONTROL POTENTIAL FOR JAPANESE KNOTWEED

Ando, Y. (1986) Seasonal prevalence in outbreaks of the Japanese beetle, *Popillia japonica* Newman (Coleoptera: Scarabaeidae). *Japanese Journal of Applied Entomology and Zoology*, **30** (2), 111-116.

Beerling, D.J. and Dawah, H.A. (1993) Abundance and diversity of invertebrates associated with *Fallopia japonica* (Houtt. Ronse Decraene) and *Impatiens glandulifera* (Royle): two alien plant species in the British Isles. *The Entomologist*, **112**, 127-139.

Child, L.E., de Waal, L.C. and Wade, P.M. (1993) Herbicides – is there a better way to control *Fallopia japonica*? In: Thomas, J.-M. (ed.), *Maitrice des Adventices par Voie Non Chimique*. July 5 – 9, Quetigny, Dijon, France. pp. 217-223.

Ellis, M.B. and Ellis, J.P. (1985) *Microfungi on land plants*. Croom Helm, London.

Emery, M.J. (1983) The ecology of Japanese knotweed (*Reynoutria japonica* Houtt.). Its herbivores and pathogens and their potential as biological control agents. Unpublished M.Sc. thesis, University of Wales, Bangor.

Fowler S.V. and Holden, A.N.G. (1994) Classical biological control for exotic invasive weeds in riparian and aquatic habitats – Practice and prospects. In: Waal, L.C. de, Child, L.E., Wade P.M. and Brock, J.H. (eds), *Ecology and Management of Invasive Riparian Plants*. Wiley, Chichester. pp. 173-182.

Fowler, S.V. and Schroeder, D. (1990) Biological control of invasive plants in the UK. Prospects and possibilities. *Conference Proceedings of the Industrial Ecology Group of the British Ecological Society*, Cardiff. pp. 130-137.

Fowler, S.V., Holden, A.N.G. and Schroeder, D. (1991) The possibilities for classical biological control of weeds of industrial and amenity land in the UK using introduced insect herbivores or plant pathogens. *Brighton Crop Protection Conference*, 1991.

Harada, Y. (1978) New hosts and biologic specialization in the aerial state *of Puccinia phragmitis* in Japan. *Transactions of the Mycological Society of Japan*, **19**, 433-438.

Harada, Y. (1987) Aerial hosts for three graminicolous *Puccinia* species (Uredinales) in Japan, with a designation of biologic forms in *Puccinia phragmitis*. *Transactions of the Mycological Society of Japan*, **28**, 197-208.

Holden, A.N.G., Fowler, S.V. and Schroeder, D. (1992) Invasive weeds of amenity land in the UK: Biological control – the neglected alternative. *Aspects of Applied Biology*, **29**, 325-332.

Shaw, R. (1997) Towards the biological control of Japanese knotweed in the United Kingdom. International Institute of Biological Control. Unpublished Report.

Strobel, G.A. (1991) Biological control of weeds. *Scientific American*, July, 50-60.

Suzuki, N. (1985) Habitat selection of three Chrysomelid species associated with *Rumex* spp. *Oecologia* (Berlin), **66**, 187-193.

Suzuki, N. (1985) Resource utilization of three Chrysomelid beetles feeding on *Rumex* plants with diverse vegetational background. *Japanese Journal of Ecology*, **35**, 225-234.

Suzuki, N. (1986) Interspecific competition and coexistence of the two Chrysomelids, *Gastrophysa atrocyanea* Motschulsky and *Galerucella vittaticollis* Baly (Coleoptera: Chrysomelidae) under limited food resource conditions. *Ecological Research*, **1**, 259-268.

Suzuki, N. (1986) Resource exploitation of larvae of *Gastrophysa atrocyanea* Motschulsky and *Galerucella vittaticollis* Baly (Coleoptera: Chrysomelidae) under a limited resource condition. *Res. Popul. Ecol.*, **28**, 69-83.

Suzuki, N. (1988) Life history characteristics and resource utilization in the Chrysomelid species associated with *Rumex* plants. *Chrysomela*, **18**, 78.

Suzuki, N. (1989) Effects of herbivory by Chrysomelid beetles on the growth and survival of *Rumex* plants. *Ecological Research*, **4**, 373-385.

Suzuki, N. (1994) Increased reproductive allocation of *Rumex japonicus* caused by leaf clipping. *Plant Species Biol.*, **9**, 31-35.

Zimmermann, K. and Topp, W. (1991) Herbivore insect community in a *Reynoutria* (Polygonaceae) hybrid zone in central Europe. *Proceedings of the 4th ECE/XIII*, Gödöllö, pp. 607-609.

Zimmermann, K. and Topp, W. (1991) Colonization of insects on introduced plants of the genus *Reynoutria* (Polygonaceae) in central Europe. *Zoologischer Jahrbücher (Systematik)* **118**, pp. 377-390.

Zwoelfer, H. (1973) Possibilities and limitations in biological control of weeds. *OEPP/EPPO Bulletin*, **3**, 19-30.

8.3 CONTROL METHODS AND TREATMENTS FOR JAPANESE KNOTWEED

Ahrens, J.F. (1963) Chemical control of Japanese Fleeceflower (*Polygonum cuspidatum* Sieb.). *Proceedings of Northeastern Weed Control Conference*, **17**. pp. 397-400.

Anon (1973) Herbicidal maleic hydrazide/dichloropropionic acid mix – for combating *Reynoutria japonica*. *Agricultural Chemistry*, **9** (2), Kumiai Chemical Ind. Co. Ltd., 3.

Baker, R.M. (1988) Mechanical control of Japanese knotweed in an SSSI. *Aspects of Applied Biology*, **16**, 189-192.

Barrett, P.R.F. (1990) Development of a long-lance sprayer. *Aquatic Weeds Research Unit Progress Report*. AFRC Institute of Arable Crops Research, Reading, UK. p.9.

Beerling, D.J. (1990) Ecology and Control of Japanese knotweed (*Reynoutria japonica* Houtt.) and Himalayan balsam (*Impatiens glandulifera. Royle*) on Riverbanks in South Wales. Unpublished Ph.D. thesis. University of Wales, Cardiff. 138 pp.

Beerling, D.J. (1990) The use of non-persistent herbicides, glyphosate and 2,4-D amine to control riparian stands of Japanese knotweed (*Reynoutria japonica* Houtt.). *Regulated Rivers, Research and Management*, **5**, 413-417.

Beerling, D.J. (1990) The use of non-persistent herbicides to control riparian stands of Japanese knotweed (*Reynoutria japonica* Houtt.). In: *Proceedings of the Conference of the Industrial Ecology Group of the British Ecological Society*, Cardiff. pp. 121-130.

Beerling, D.J. (1991) The testing of cellular concrete revetment blocks resistant to growths of *Reynoutria japonica* Houtt. (Japanese knotweed). *Water Research*, **25**, 495-498.

Bing, A. (1977) Glyphosate to control perennial weeds in landscape plantings. *Proceedings Northeastern Weed Science Society*, **31**, p. 327.

Briggs, M. (1991) Management of churchyards. *BSBI News*, **58**, 43-44.

Child, L.E. (1999) Vegetative regeneration and distribution of *Fallopia japonica* and *Fallopia* x *bohemica*: Implications for control and management. Unpublished Ph.D. thesis, Loughbrough University, UK.

Child, L.E. and Waal, L.C. de (1997) The use of Geographical Information Systems in the management of *Fallopia japonica* in the urban environment. In: Brock, J.H., Wade, P.M., Pysek, P. and Green, D. (eds), *Plant Invasions: Studies from North America and Europe*. Backhuys, Leiden. pp. 207-220.

Child, L.E. and Wade, P.M. (1997) Reasons for the succesful invasion of *Fallopia japonica* in the British Isles and implications for management. In: National Institute of Agro-Environmental Sciences, *Proceedings International workshop on biological invasions of ecosystem by pests and beneficial organisms*, February 25-27, 1997, Tsukuba, Japan. pp. 253-267.

Child, L.E., de Waal, L.C. and Wade, P.M. (1993) Herbicides – is there a better way to control *Fallopia japonica*? In: Thomas, J.-M. (ed.), *Maitrice des Adventices par Voie Non Chimique*. July 5 – 9, Quetigny, Dijon, France. pp. 217-223.

Child, L.E., Wade, P.M. and Wagner, M. (1998) Cost effective control of *Fallopia japonica* using combination treatments. In: Starfinger, U., Edwards, K., Kowarik, I. and Williamson, M. (eds), *Plant Invasions: Ecological Mechanisms and Human Responses*. Backhuys, Leiden. pp. 143-154.

Child, L.E., Waal, L.C. de, Wade, P.M. and Palmer, J.P. (1992) Control and management of *Reynoutria* species (Knotweed). *Aspects of Applied Biology*, **29**, 295-307.

Cooke, A.S. (1988) Japanese knotweed: its status as a pest and its control in nature conservation areas. Unpublished Report, English Nature, Peterborough.

Cooke, A.S. (1991) The use of herbicides on National Nature Reserves. *Brighton Crop Protection Conference-Weeds 1991*, **5C-1**, pp. 619-626.

Cyanamid (no date) *Arsenal herbicide.* Technical dossier, American Cyanamid Company, Princeton, New Jersey.

de Waal, L.C. (1995) Treatment of *Fallopia japonica* near water – a case study. In: Pysek, P., Prach, K., Rejm·nek, M. and Wade, P.M. (eds), *Plant Invasions – General Aspects and Special Problems.* SPB Academic Publishing, Amsterdam. pp. 203-212.

de Waal, L.C., Child, L.E. and Wade, P.M. (1995) The management of three alien invasive riparian plants: *Impatiens glandulifera* (Himalayan balsam), *Heracleum mantegazzianum* (giant hogweed) and *Fallopia japonica* (Japanese knotweed) In: Harper, D. and Ferguson, A. (eds), *The Ecological Basis of River Management.* Wiley, Chichester. pp. 315-321.

Department of the Environment (1991) *Waste Management – The Duty of Care, a Code of Practice. Environmental Protection Act 1990.* HMSO, London.

Edmonds, J.L. (1986) The distribution and control of an introduced species *Reynoutria japonica* (Japanese knotweed) in selected rivers in South West Wales. Unpublished Report, Welsh Water.

Environment Agency (1996) *Guidance for the control of invasive plants near watercourses.* Environment Agency, Bristol.

Environment Agency and Cornwall County Council (1998) *Japanese knotweed. How to control it and prevent its spread.* Environment Agency, Bodmin.

Ferron, M. (1966) Essais de désherbages chimique des gazons. *Recherches Agronomiques*, **11**, 1-38.

Figueroa, P.F. (1989) Japanese knotweed screening trial applied as a roadside spray. *Proceedings Western Society of Weed Science*, **42**, 288-298.

Gregson, D. (1981) Control of Japanese knotweed in Glamorgan. *Conservation Officers Bulletin*, **7**.

Gritten, R.H. (1990) The control of invasive plants in the Snowdonia National Park. *Proceedings of the Conference of the Industrial Ecology Group of the British Ecological Society*, Cardiff. pp. 80-85.

Groves, R.H. (1989) Ecological control of invasive terrestrial plants. In: Drake, J.A., Mooney, H.A., di Castri, F., Groves, R.H., Kruger, F.J., Rejm·nek, M. and Williamson, M. (eds), *Biological Invasions: a Global Perspective.* SCOPE 37. Wiley, Chichester. pp. 437-461.

Gunn, I.D.M. (1986) Biology and control of Japanese knotweed (*Reynoutria japonica*) and Himalayan balsam (*Impatiens glandulifera*) on river banks. Unpublished M.Sc. thesis, University of Wales, Cardiff.

Harper, C.W. and Stott, K.G. (1966) Chemical control of Japanese knotweed. *Proceedings 8th British Weed Control Conference*, pp. 511-515.

Hawke, C. and Williamson, D.R. (1995) *Japanese Knotweed in Amenity Areas*. Arboricultural Research Note 106. Arboricultural Advisory and Information Service, Farnham, Surrey.

Hill, D.J. (1994) A practical strategy for the control of *Fallopia japonica* (Japanese knotweed) in Swansea and the surrounding area, Wales. In: Waal, L.C. de, Child, L.E., Wade, P.M. and Brock, J.H. (eds), *Ecology and Management of Invasive Riverside Plants*. Wiley, Chichester. pp. 195-198.

Ito, M., Ueki, K. and Sakamoto, S. (1982) Studies on the total vegetation control in railroads. I. Major weeds and factors affecting their distribution. *Weed Research, Japan*, **27** (1), 41-47.

Ivens, G.W. (1993) *The UK Pesticide Guide*. CAB International, University Press, Cambridge.

Jackson, P. and Turtle, C. (1986) *The biology and control of Japanese knotweed (Reynoutria japonica) with particular reference to Telford*. Telford Nature Conservation Project, Stirchley Grange.

Jennings, V.M. and Fawcett. R.S. (1980) *Weed Control: Japanese Polygonum (Polygonum cuspidatum* Sieb. and Zucc.). Co-operative Extension service E5. Iowa State University, Ames, Iowa.

Kluge, R.L., Zimmermann, H.G., Cilliers, C.J. and Harding, G.B. (1986) Integrated control for invasive alien weeds. In: Macdonald, I.A.W., Kruger, F.J. and Ferrar, A.A. (eds), *The Ecology and Management of Biological Invasions in Southern Africa*. Oxford University Press, Cape Town.

Locandro, R.R. (1978) Weed Watch: Japanese Bamboo, 1978. *Weeds Today*, **9**, 21-22.

Lynn, L.B., Rogers, R.A. and Graham, J.C. (1979) Response of woody species to glyphosate in northeastern states. *Proceedings Northeastern Weed Science Society*, **33**. pp. 336-342.

Meade, J.A. and Locandro, R.R. (1979) Japanese bamboo. *The Journal*, 1 – 4.

Ministry of Agriculture, Fisheries and Food (1995) *Guidelines for the use of herbicides on weeds in or near watercourses and lakes*. MAFF Publications, London.

Murphy, K.J. and Barrett, P.R.F. (1990) Chemical control of aquatic weeds. In: Pieterse, A.H. and Murphy, K.J. (eds), *Aquatic Weeds – the Ecology and Management of Nuisance Aquatic Vegetation*. Oxford Science Publications, New York.

Nashiki, M., Nomoto, T., Megure, R. and Sato, K. (1986) Effect of natural conditions and management of pastures on weed invasion in cooperative livestock. *Weed Research*, Japan, **31** (3), 221-227.

Palmer, C.G.. Reid, D.F. Godding. S.J. (1988) A review of forestry trials with a formulation of triclopyr, dicamba and 2, 4-D. *Aspects of Applied Biology*, **16**, 207-214.

Pridham, A.M.S., Schwartzbeck, R.A. and Cozart, E.R. (1966) Control of emigrant Asian perennials. *Biokemia*, **11**, 68.

Pycraft, D. (1992) Japanese knotweed – Some notes on control. *Horticultural Advisory Leaflet, Hort.*, **21**, 2, Royal Horticultural Society, Wisley, Surrey.

Richards, Moorehead and Laing Ltd. (1990) *Japanese knotweed (Reynoutria japonica) in Wales*. 1, Main Text. Director of Land Reclamation, Welsh Development Agency, Cardiff.

Richards, Moorehead and Laing Ltd. (1990) *Japanese knotweed (Reynoutria japonica) in Wales*. 2, Appendix. Director of Land Reclamation, Welsh Development Agency, Cardiff.

Roblin, E. (1988) Chemical control of Japanese knotweed (Reynoutria japonica) on river banks in South Wales. *Aspects of Applied Biology*, **16**, 201-206.

Scott, R. (1988) A review of Japanese knotweed control. Unpublished Report to Nature Conservancy Council.

Scott, R. and Marrs, R.H. (1984) Impact of Japanese knotweed and methods of control. *Aspects of Applied Biology*, **5**, 291-296.

Seiger, L.A. (1993) The ecology and control of *Reynoutria japonica* (*Polygonum cuspidatum*). Unpublished Ph.D. thesis, The George Washington University, USA.

Seiger, L.A. (1997) The status of *Fallopia japonica* (*Reynoutria japonica; Polygonum cuspidatum*) in North America. In: Brock, J.H., Wade, P.M., Pysek, P. and Green, D. (eds), *Plant Invasions: Studies from North America and Europe*. Backhuys, Leiden. pp. 95-102.

Seiger, L.A. and Merchant, H.C. (1990) The ecology and control of *Polygonum cuspidatum*. *Bulletin of the Ecological Society of America*, **71**, 322.

Seiger, L.A. and Merchant, H.C. (1997) Mechanical control of Japanese knotweed (*Fallopia japonica* Houtt. Ronse Decraene): Effects of cutting regime on rhizomatous reserves. *Natural Areas Journal*, **17**, 341-345.

Stensones, A and Garnett, R.P. (1994) Controlling invasive weeds using glyphosate. In: Waal, L.C. de, Child, L.E., Wade, P.M. and Brock, J.H. (eds), *Ecology and Management of Invasive Riparian Plants*. Wiley, Chichester. pp. 183-188.

Upchurch, R.P., Keaton, J. A. and Selman, F. L. (1965) The utilization of plant growth control substances in the maintenance of highway right-of-way and highway facilities. School of Engineering. *North Carolina State University Report 0388*. p. 124.

Wade, P.M., de Waal, L.C., Child, L.E., Dodd, F.S. and Darby, E.J. (1994) *Control of Invasive Riparian and Aquatic Weeds*. National Rivers Authority, R&D Project Record 294/7/W.

Welsh Development Agency (1994) *Model Tender Specifications for the Eradication of Japanese knotweed*. Welsh Development Agency, Cardiff.

Welsh Development Agency (1998) *The Control of Japanese Knotweed in Construction and Landscape Contracts: Model Specification*. Welsh Development Agency, Cardiff.

Welsh Development Agency (1998) *The Eradication of Japanese Knotweed: Model Tender Document*. Welsh Development Agency, Cardiff.

Young, R.G. and Aharrah, E.C. (1984) Natural control of Japanese Fleeceflower (*Polygonum cuspidatum*) on mined land. *Proceedings Symposium on Surface Mining, Hydrology, Sedimentology and Reclamation*, University of Kentucky, USA. pp. 475-478.

Zeneca (1994) *Technical Dossier*. Zeneca Crop Protection, Surrey.

8.4 DISTRIBUTION OF JAPANESE KNOTWEED AND RELATED SPECIES

Bailey, J.P., Child, L.E. and Conolly, A.P. (1996) A survey of the distribution of *Fallopia x bohemica* (Chrtek and Chrtková) J. Bailey (Polygonaceae) in the British Isles. *Watsonia*, **21**, 187-198.

Bailey, J.P., Child, L.E. and Wade, P.M. (1995) Assessment of the genetic variation and spread of British populations of *Fallopia japonica* and its hybrid *Fallopia x bohemica*. In: Pysek, P., Prach, K., Rejmánek, M. and Wade, P.M. (eds), *Plant Invasions: General Aspects and Special Problems*. SPB Academic Publishing, Amsterdam. pp. 141-150.

Beerling, D.J. (1991) The effect of riparian land use on the occurrence and abundance of Japanese knotweed *Reynoutria japonica* on selected rivers in South Wales. *Biological Conservation*, **55**, 329-337.

Beerling, D.J. and Palmer, J.P. (1994) Status of *Fallopia japonica* (Japanese knotweed) in Wales. In: Waal, L.C. de, Child, L.E., Wade P.M. and Brock, J.H. (eds), *Ecology and Management of Invasive Riparian Plants*. Wiley, Chichester. pp. 199-211.

Beerling, D.J. and Woodward. F.I. (1994) Climate change and the British scene. *Journal of Ecology*, **82**, 391-397.

Beerling, D.J., Huntley, B. and Bailey, J.P. (1995) Climate and the distribution of *Fallopia japonica*: use of an introduced species to test the predictive capacity of response surfaces. *Journal of Vegetation Science*, **6**, 269-282.

Brandes, D. (1995) Die uferflora im bereich des lago Maggiore. *Floristische Rundbriefe*, Bochum, **29** (2), 194-197.

Brandes, D. (1995) The flora of old town centres in Europe. In: *Urban Ecology as the Basis of Urban Planning*. SPB Academic Publishing, Amsterdam. pp. 49-58.

Brandes, D. and Oppermann. W. (1994) Die uferflora der oberen Weser. The urban flora of the upper part of the river Weser (Germany). *Braunschw. Naturkdl. Schr.*, **4** (3), 575-607.

Brandes, D. and Sander, C. (1995) Die vegetation von ufermauern und uferplasterungen an der Elbe – The vegetation of quays and pavements of the banks of the R. Elbe. *Braunschw. Naturkdl. Schr.*, **4**, 899-912.

Charter, J.R. (1997) The spread of Japanese knotweed and Himalayan balsam, invasive alien weeds in Chesterfield, Derbyshire. *BSBI News*, **75**, 51-54.

Chrtek, J. and Chrtkov·, A. (1983) *Reynoutria* x *bohemica*, Novy Krieznec Z celedi Rdesnovitych. *J. Nat. Mus. Praha Hist. Nat.*, **152**, 120 (in Czech).

Conolly, A.P. (1977) The distribution and history in the British Isles of some alien species of *Polygonum and Reynoutria. Watsonia*, **11**, 291-311.

Conolly, A.P. (1998) *Fallopia* x *bohemica* – A new record from Australia? *BSBI News*, **78**, 88.

Dickson, J.H. (1994) A large stand of giant knotweed (*Fallopia sachalinensis*) at Skipness, Kintyre. *Glasgow Naturalist*, **22** (4), 421-422.

Dickson, J.H. and Watson. K. (1994) *Fallopia* x *bohemica* (Chrtek and Chrtkova) J. Bailey in the Glasgow area. *Glasgow Naturalist*, **22** (4), 423.

Duvigneaud, J. and Saintenoy-Simon, J. (1989) Quelques observations floristiques effectuées à l'occasion de deux excursions à la Heid des Gattes. *Institut de Aoristique Belgo-Luxembourgeois ashl.*, **7** (3), 35-38.

Edmonds, J.L. (1986) The distribution and control of an introduced species *Reynoutria japonica* (Japanese knotweed) in selected rivers in South West Wales. Unpublished Report, Welsh Water.

Ellenberg, H. (1978) *Vegetation Mitteleuropas mit den Alpen in ökologischer Sicht*. Ulmer, Stuttgart.

Fremstad, E. and Elven, R. (1997) Fremmede planter i Norge. De store *Fallopia* arten. *Blyttia*, **55**, 3-14.

Gunn, I.D.M. (1986) Biology and control of Japanese knotweed (*Reynoutria japonica*) and Himalayan balsam (*Impatiens glandulifera*) on river banks. Unpublished M.Sc. thesis, University of Wales, Cardiff.

Izco, J. (1974) *Reynoutria japonica* Houtt. en Espana. *Boletin Real Socieded Espanola de Historia Natural*, **72**, 25-28.

Kanai, H. (1991) Distribution of popular plants in Miyagi Prefecture, North Japan. *Journal of Japanese Botany*, **66** (2), 83-110.

Kanai, H. (1992) Distribution of popular plants in Aichi Prefecture, Central Japan. *Bull. Natn. Sci. Mus. Tokyo, Ser. B*, **18** (2), 59-81.

Kanai, H. (1992) Distribution of popular plants in Fukui Prefecture, Central Japan. *J. Jpn. Bot.*, **67**, 291-309.

Kanai, H. (1992) Distribution of popular plants in Hyogo Prefecture, West Japan. *Bull. Natn. Sci. Mus.. Tokyo, Ser. B*, **18** (4), 149-166.

Kanai, H. (1992) Distribution of popular plants in Nara Prefecture, Western Japan. *Bull. Natn. Sci. Mus. Tokyo, Ser. B*, **19** (4), 137-155.

Kanai, H. (1992) Distribution of popular plants in Okayama Prefecture, West Japan. *Journal of Japanese Botany*, **67**, 347-364.

Kanai, H. (1993) Distribution of popular plants in Gifu Prefecture, Central Japan. *Bull. Natn. Sci. Mus., Tokyo, Ser. B*, **19** (2), 59-78.

Kanai, H. (1994) Distribution of popular plants in Kumamoto Prefecture, Western Japan. *Bull. Natn. Sci. Mus. Tokyo, Ser. B*, **20** (4), 163-180.

Kanai, H. (1995) Distribution of popular plants in Mie Prefecture, Central Japan. *Journal of Japanese Botany*, **70**, 154-172.

Kanai, H. (1995) Distribution of popular plants in Shiga Prefecture, Central Japan. *Bull. Natn. Sci. Mus., Tokyo, Ser. B*, **21** (3), 131-150.

Kanai, H. (1996) Distribution of popular plants in Chubu District, Central Japan. *Journal of Japanese Botany*, **71** (6), 338-354.

Kanai, H. (1996) Distribution of popular plants in Gunma Prefecture, Central Japan. *Journal of Japanese Botany*, **71**, 125-144.

Kanai, H. (1996) Distribution of popular plants in Saga Prefecture, Western Japan, *Journal of Japanese Botany*, **71**, 29-38.

Kanai, H. and Konta, F. (1987) Distribution of commonly occurring plants in Shizuoka Prefecture, Central Japan. *Bulletin of the National Science Museum, Japan B*, **13** (4), 151-170.

Komzha, A.L. and Popov, K.P. (1990) New data on the adventive flora of the North Ossetian Autonomous Soviet Socialist Republic. *Bot. Zh. (Leningr.)*, **75**, 108-110.

Landolt, E. (1991) Distribution patterns of flowering plants in the city of Zurich. In: Esser G. and Overdieck, D. (eds), *Modern Ecology: Basic and Applied Aspects*. Elsevier, Amsterdam. pp. 807-822.

Locandro, R.R. (1975) Distribution, phenological and morphological characteristics of Japanese knotweed. *Abstracts of 1975 meeting of the Weed Science Society of America*, February 4-7, 1975.

Locandro, R.R. (1979) Updating information for Japanese bamboo (*Polygonum cuspidatum* Sieb. & Zucc.) in the United States and Europe. Abstract of 1979 Meeting of the Weed Science Society of America. New Jersey, USA. p. 170.

Locandro, R.R. (1984) The distribution of *Polygonum cuspidatum* Sieb. and Zucc. in Western Europe. *Proceedings of the 7th International Colloquium on Ecology*. Paris, France. pp. 133-137.

Lohmeyer, W. and Sukopp. H. (1992) Agriophyten in der vegetation Mittleuropas. *Schr. Reihe Vegetationskde*. **25**. Bonn-Bad Godesberg.

Macpherson, E.L.S. and Macpherson. P. (1975) Alien *Polygonum* spp. *Glasgow Naturalist*, **19** (3), 203-204.

Masalles, R.M., Sans. F.X., Pino. J. and Chamorro, L. (1996) Contribution to the knowledge of the syanthropic flora of Catalonia (NE Iberian Peninsula). *Folia Botanica Miscellanea*, **10**, 77-84.

Mennema, J., Quene-Boterenbrood, A.J. and Plate, C.L. (1985) *Atlas van de Nederlandse flora*. Bohn, Scheltema and Holkema, Utrecht.

Numata. N. (1974) *The Present Flora: Its General Features and Regional Divisions. Flora and Vegetation of Japan*. Elsevier, Amsterdam.

Ohba, T. (1969) Eine pflanzensoziologische Gliederung über die Wustepflanzengesellschaften auf alpinen Stufen Japans. *Bulletin of the Kanagawa Prefectural Museum*, **1** (2), 23-70.

Ohwi, J. (1965) *Flora of Japan*. Smithsonian Institution, Washington, DC.

Patterson, D. T . (1976) The history and distribution of five exotic weeds in N. Carolina. *Castanea*, **41** (2), 177-180.

Perring, F.H. and Walters, S.M. (1990) *Atlas of the British Flora*. Botanical Society of the British Isles, London.

Schmitz, J. and Strank, K.J. (1985) Die drei *Reynoutria*-Sippen (*Polygonaceae*) des Aachener Stadtwaldes. *Gött. Flor. Rundbr.*, **19**, 1725.

Schmitz, J. and Strank, K.J. (1986) Zur sociologie der *Reynoutria*-Sippen (Polygonaceae) im Aachener Stadtwaldes. *Decheniana* (Bonn) **139**, 141-147.

Schnitzler, A. and Muller, S. (1998) Ecologie et biogéographie des plantes haute-ment invasives en Europe: Les renouées géantes du Japon (*Fallopia japonica* et *F. sachalinenesis*). *Rev. Ecol. (Terre Vie)*, **53**, 3-37 (in French, English Summary).

Stypinski, P. (1977) New localities of *Polygonum sachalinense* Schm. and *P. cuspida-tum* Sieb. et Zucc. in Varmia and Mazuria (NE Poland). *Fragmenta Floristica* et *Geobotanica*, **23**, part 1, 316.

Takenaka, A., Washitani, I., Kuramoto, N. and Inoue, K. (1996) Life history and demographic features of *Aster kantoensis*. An endangered local endemic of floodplains. *Biological Conservation*, **78**, 345-352.

Van Rompaey, E. and Delvosalle, L. (1979) *Atlas de la Flore Belge et Luxembourgeoise: Pteridophytes et Spermatophytes*. 2nd ed. Jardin Botanique National de Belgique.

Visnak, R. (1986) A contribution to the knowledge of anthropogenic vegetation in northern Bohemia. Czechoslovakia especially in the town of Liberec. *Preslia* (Prague), **58** (4), 353-368.

Wade A.E., Kay, Q.O.N., Ellis, R.G. and The National Museum of Wales (1994) *Flora of Glamorgan*. HMSO, London.

8.5 ECOLOGY OF JAPANESE KNOTWEED

Adachi, N., Terashima, I. and Takahashi, M. (1996) Central die-back of monoclon-al stands of *Reynoutria japonica* in an early stage of primary succession on Mount Fuji. *Annals of Botany*, **77**, 477-486.

Adachi, N., Terashima, I. and Takahashi, M. (1996) Mechanisms of central die-back of *Reynoutria japonica* in the volcanic desert on Mt. Fuji: A stochastic model analysis of rhizome growth. *Annals of Botany*, **78**, 169-179.

Adachi, N., Terashima, I. and Takahashi, M. (1996) Nitrogen translocation via rhi-zome systems in monoclonal stands of *Reynoutria japonica* in an oligotrophic desert on Mt. Fuji: Field experiments. *Ecological Research*, **11**, 175-186.

Adler, L. (1993) Zur strategie und Vergesellschaftung des neophyten *Polygonum cuspidatum* unter besonderer berucksichtigung der mahad. *Tuexenia*, **13**, 373-397.

Alberternst, B. (1995) *Handbuch Wasser 2: Kontrolle des Japan-Knöterichs an Flieflgewßssern. II. Untersuchungen zu biologie und ökologie der neophytischen Knöterich-Arten.* Landesenstalt für Umweltschultz Baden-Württemberg, Karlsruhe.

Alberternst, B., Konold, W. & Böcker. R. (1995) Untersuchungen zur Morphologie und Blutenbiologie bei der Gattung Reynoutria. *Ber. Inst Landschafts - Pflanzenökologie Univ. Hohenheim*, **4**, 141-150.

Anon (1996) Japanese Knotweed. *British Wildlife*, **7**, 175.

Beerling, D.J., Bailey, J.P. and Conolly, A.P. (1994) Biological Flora of the British Isles – *Fallopia japonica* (Houtt.) Ronse Decraene. *Journal of Ecology*, **82**, 959-979.

Berney, M. (1971) The Japanese knotweed. *Rev. Hort. Suisse*, **44** (5), 138-139.

Brock, J.H. (1994) Technical note: Standing crop of *Fallopia japonica* in the autumn of 1991 in the United Kingdom. *Preslia*, **66**, 337-343.

Brock, J.H. and Wade, P.M. (1992) Regeneration of *Fallopia japonica*, Japanese knotweed, from rhizome and stems: Observations from greenhouse trials. *IXe Colloque International sur la Biologie des Mauvaises Herbes*. Dijon (France). pp. 85-94.

Brock, J.H., Child, L.E., de Waal, L.C. and Wade, P.M. (1995) The invasive nature of *Fallopia japonica* is enhanced by vegetative regeneration from stem tissues. In: Pysek, P., Prach, K., Rejmánek, M. and Wade, P.M. (eds), *Plant Invasions: General Aspects and Special Problems*. SPB Academic Publishing, Amsterdam. pp. 131-139.

Chiba, N. and Hirose, T. (1993) Nitrogen acquisition and use in three perennials in the early stage of primary succession. *Functional Ecology*, **7**, 287-292.

Emery, M.J. (1983) The ecology of Japanese knotweed (*Reynoutria japonica* Houtt.). Its herbivores and pathogens and their potential as biological control agents. Unpublished M.Sc. thesis, University of Wales, Bangor.

Ferron, M. (1966) Essais de désherbages chimiques des gazons. *Recherches Agronomiques*, **11**, p38.

Ferron, M. (1968) Studies of the Japanese Polygonum-D. *Recherches Agronomiques*, **13**, 56.

Figueroa, P. F. (1988) *Literature review: Japanese knotweed (Polygonum cuspidatum), a potential noxious weed.* Weyerhaeuser Research Report. Weyerhaeuser & Centralia Research Institute, Centralia, Washington.

Flower, C.J. (1987) Biology and control of Japanese knotweed (*Reynoutria japonica*) on river banks. Unpublished report, Welsh Water.

Galle, P. (1977) Untersuchungen zur Blutenentwicklung der Polygonaceen. *Bot. Jahrb. Syst.*, **98** (4), 449-489.

Gilbert, O. (1989) *The Ecology of Urban Habitats.* Chapman and Hall, London.

Grime, J.P., Hodgson, J.G. and Hunt, R. (1988) *Comparative Plant Ecology.* Unwin Hyman, London.

Hara, T. (1994) Growth and competition in clonal plants – persistence of shoot populations and species diversity. *Folia Geobot. Phytotax.*, **29**, 181-201.

Harranger, J. (1985) Informations générales les renouées bambous. *Phytoma*, **5**, 366.

Hawke, C. and Williamson, D.R. (1995) *Japanese Knotweed in Amenity Areas.* Arboricultural Research Note 106. Arboricultural Advisory and Information Service, Farnham, Surrey.

Hirose, T. (1984) Nitrogen use efficiency in growth of *Polygonum cuspidatum* Sieb. et Zucc. *Annals of Botany*, **54**, 695-704.

Hirose, T. (1986) Nitrogen uptake and plant growth. 2. An empirical model of vegetative growth and partitioning. *Annals of Botany*, **58**, 487-496.

Hirose, T. (1987) A vegetative plant growth model: adaptive significance of phenotypic plasticity in matter partitioning. *Functional Ecology*, **1**, 195-202.

Hirose, T. (1988) Modelling the relative growth rate as a function of plant nitrogen concentration. *Physiologia Plantarium*, **72**, 185-189.

Hirose, T. and Kitajima, K. (1986) Nitrogen uptake and plant growth I. Effect of nitrogen removal on growth of *Polygonum cuspidatum*. *Annals of Botany*, **58**, 479-486.

Hirose, T. and Tateno, M. (1984) Soil nitrogen patterns induced by colonisation of *Polygonum cuspidatum* on Mt. Fuji. *Oecologia*, **61**, 218-223.

Horn, P. (1997) Seasonal dynamics of aerial biomass of *Fallopia japonica*. In: Brock, J.H., Wade, P.M., Pysek, P. and Green, D. (eds), *Plant Invasions: Studies from North America and Europe*. Backhuys, Leiden. pp. 203 - 206.

Imahara, H., Hatayama, T., Kuroda, S., Horie, Y., Inoue, E., Wakatsuki, T., Kitamara, T., Fujimoto, S., Ohara, A. and Hashimoto, K. (1992) Production of phytochelatins in *Polygonum cuspidatum* on exposure to copper but not to zinc. J. *Pharmacobio-Dyn*, **15** (12), 667-671.

Impens, R., M'Vunzu. Z. et Nangniot, P. (1972) Determination du plomb sur les végétaux croissant en bordure des autoroutes. *Environmental Health Aspects of Lead. Proc. Int. Symp.* Amsterdam. pp. 35-143.

Jackson, P. and Turtle, C. (1986) *The biology and Control of Japanese Knotweed (Reynoutria japonica) with Particular Reference to Telford.* Telford Nature Conservation Project, Stirchley Grange.

Jenik, J. (1994) Clonal growth in woody plants: a review. *Folia Geobot. Phytotax., Praha*, **29**, 291-306.

Krause, A. (1983) Zur entwicklung des seifenkraut-queckenrasens (*Saponaria officinalis-Agropyron repens*-Gesellschaft) in Mundungsgebiet der Ahr. *Decheniana*, **136**, 20-29.

Krause, A. (1990) Neophyten an der Ahr Stand der Ausbreitung 1988. *Tuexenia*, **10**, 49-55.

Kubo, H., Nozue, M., Kawasaki, K. and Yasuda, H. (1995) Intravacuolar spherical bodies in *Polygonum cuspidatum*. *Plant Cell Physiol.*, **36** (8), 1453-1458.

Kretz, M. (1992) *Kontrollmethoden zur Eindammung von* Reynoutria japonica. Zwischenbericht. Report for WBA Offenburg.

Kubota, K., Nishizono, H., Suzuki, S. and Ishii, F. (1988) A copper-binding protein in root cytoplasm of *Polygonum cuspidatum* growing in a metalliferous habitat. *Plant Cell Physiology*, **26**, 1029-1033.

Locandro, R.R. (1973) Reproduction ecology of *Polygonum cuspidatum*. Unpublished Ph.D. thesis, Rutgers University, NJ, USA.

Locandro, R.R. (1975) Distribution, phenological and morphological characteristics of Japanese knotweed. *Abstracts of 1975 meeting of the Weed Science Society of America*, February 4-7, 1975.

Locandro, R.R. (1975) Ecological reproduction of Japanese knotweed. *Abstracts of 1975 meeting of the Weed Science Society of America*, February 4-7, 1975.

Locandro, R.R. (1978) Weed Watch: Japanese Bamboo, 1978. *Weeds Today*, **9**, 21-22.

Manko, Y.I. (1980) Volcanism and the dynamics of vegetation. *Bot. Zh. (Leningrad)*, **65**, 457-469.

Mariko, S. and Kachi, N. (1995) Seed ecology of *Lobelia bobinensis* Koidz. (Campanulaceae), an endemic species in the Bonin Islands (Japan). *Plant Species Biol.*, **10**, 103-110.

Mariko, S. and Koizumi, H. (1993) Respiration for maintenance and growth in *Reynoutria japonica* ecotypes from different altitudes on Mt. Fuji. *Ecological Research*, **8** (2), 241-246.

Mariko, S., Koizumi, H., Suzuki, J-I and Furukawa, A. (1993) Altitudinal variations in germination and growth responses of *Reynoutria japonica* populations on Mt Fuji to a controlled thermal environment. *Ecological Research*, **8**, 27-34.

Maruta, E. (1976) Seedling establishment of *Polygonum cuspidatum* on Mt. Fuji. *Japanese Journal of Ecology*, **26**, 101-105.

Maruta, E. (1981) Size structure in *Polygonum cuspidatum* on Mt. Fuji. *Japanese Journal of Ecology*, **31**, 441-445.

Maruta, E. (1983) Growth and survival of current-year seedlings of *Polygonum cuspidatum* at the upper distributional limit on Mt. Fuji. *Oecologia*, **60**, 316-320.

Maruta, E. (1994) Seedling establishment of *Polygonum cuspidatum* and *Polygonum weyrichii* var. *alpinum* at high altitudes of Mt. Fuji. *Ecological Research*, **9**, 205-213.

Maruta, E. and Saeki, T. (1976) Transpiration and leaf temperature of *Polygonum cuspidatum* on Mt. Fuji. *Japanese Journal of Ecology*, **26**, 25-35.

Masuzawa, T. (1985) Ecological studies on the timberline of Mt. Fuji. I. Structure of plant community and soil development on the timberline. *Bot. Mag. Tokyo*, **98**, 15-28.

Meade, J.A. and Locandro, R.R. (1979) Japanese bamboo. *The Journal*, 1-4.

Muller, N. (1995) River dynamics and floodplain vegetation and their alterations due to human impact. *Arch. Hydrobiol. Suppl.*, **314**, 477-512.

Nakamura, T. (1984) Vegetational recovery of landslide scars in the upper reaches of the Oi River, Central Japan. *Journal of the Japanese Forestry Society*, **66** (8), 328-332.

Natori, T. and Totsuka, T. (1984) An evaluation of high resistance in *Polygonum cuspidatum* to sulfur dioxide (SO_2). *Japanese Journal of Ecology*, **34**, 153-159.

Natori, T. and Totsuka, T. (1988) Responses of dry weight growth under SO_2 stress in an SO_2-tolerant plant, *Polygonum cuspidatum*. *Ecological Research*, **3**, 1-8.

Natori, T. and Totsuka, T. (1985) Studies on evaluations of absorption capacity of air pollutants by plant population. Effects of light intensity on stomatal resistance of *Polygonum cuspidatum* during sulfur dioxide fumigation. *Kokuritsu Kogai Kenkyusho Kenkyu Hokoku*, **82**, 29-37.

Nishida, T. (1989) Is lifetime data always necessary for evaluating the "intensity" of selection? *Evolution*, **43** (8), 1826-1827.

Nishitani, S. and Masuzawa, T. (1996) Germination characteristics of two species of *Polygonum* in relation to their altitudinal distribution on Mt. Fuji, Japan. *Arctic and Alpine Research*, **28** (1), 104-110.

Nishizono, H., Kubota, K., Suzuki, S. and Ishii. F. (1989) Accumulation of heavy metals in cell walls of *Polygonum cuspidatum* roots from metalliferous habitats. *Plant Cell Physiol.*, **30** (4), 595-598.

Paal, J. (1994) The tall herb communities in the Far East. *Svensk Botanisk Tidskrift*, **88** (4), 221-226.

Palmer, J.P. (1990) Japanese knotweed (*Reynoutria japonica*) in Wales. In: *Conference of Industrial Ecology Group of the British Ecological Society.* Cardiff. pp. 80-85.

Prach, K. and Wade, P.M. (1992) Population characteristics of expansive perennial herbs. *Preslia*, **64**, 45-51.

Pridham, A.M.S. and Bing, A. (1975) Japanese-Bamboo. *Plants Gard.*, **31** (2), 56-57.

Richards, Moorehead and Laing Ltd. (1990) *Japanese knotweed (Reynoutria japonica) in Wales.* 1, Main Text. Director of Land Reclamation, Welsh Development Agency, Cardiff.

Richards, Moorehead and Laing Ltd. (1990) *Japanese knotweed (Reynoutria japonica) in Wales.* 2, Appendix. Director of Land Reclamation, Welsh Development Agency, Cardiff.

Schmitz, J. and Strank, K.J. (1985) Die drei *Reynoutria*-Sippen (Polygonaceae) des Aachener Stadtwaldes. *Gött. Flor. Rundbr.*, **19**, 1725.

Schmitz, J. and Strank, K.J. (1986) Zur sociologie der *Reynoutria*-Sippen (Polygonaceae) im Aachener Stadtwaldes. *Decheniana* (Bonn) **139**, 141-147.

Schnitzler, A. and Muller, S. (1998) Ecologie et biogéographie des plantes hautement invasives en Europe: Les renouées géantes du Japon (*Fallopia japonica* et *F. sachalinenesis*). *Rev. Ecol. (Terre Vie)*, **53**, 3-37 (in French, English Summary).

Schuldes, H. and Kubler, R. (1991) Neophyten als Problempflanzen im Naturschutz. *Arbeitsbl. Naturschutz.*, **12**, 116.

Schuldes, H. and Kubler, R. (1990) Ökologie und Vergesellschaftung von *Solidago candensis* et *gigantea, Reynoutria japonica* et *sachalinense, Impatiens glandulifera, Helianthus tuberosus, Heracleum mantegazzianum.* Ihre Verbreitung in Baden-Wurttemberg sowie Notwendigkeit und Moglichkeiten ihrer Bekampfung. Studie im Auftrag des Ministeriums für Umwelt Baden-Wurttemberg.

Schwabe, A. and Kratochwil, A. (1991) Gewasser-begleitende Neophyten und ihre Beurteilung aus Naturschutz-Sicht unter besonderer Berucksichtigung Sudwestdeutschlands. *NNA-Berichte,* **4** (1).

Seiger, L.A. (1993) *The ecology and control of Reynoutria japonica (Polygonum cuspidatum).* Unpublished Ph.D. thesis, The George Washington University, USA.

Seiger, L.A. (1997) The status of *Fallopia japonica (Reynoutria japonica; Polygonum cuspidatum)* in North America. In: Brock, J.H., Wade, P.M., Pysek, P. and Green, D.P. (eds), *Plant Invasions: Studies from North America and Europe.* Backhuys, Leiden. pp. 95-102.

Seiger, L.A. and Merchant, H.C. (1990) The ecology and control of *Polygonum cuspidatum. Bull. Ecol. Soc. Am.,* **71**, 322.

Shibata, O. and Arai, T. (1970) Seed germination in *Polygonum reynoutria* Makino grown at different altitudes. *Japanese Journal of Ecology,* **20** (1), 9-11.

Shibata, O., Arai, T. and Kinoshita, T. (1975) Photosynthesis in *Polygonum reynoutria* L. ssp. *asiatica* grown at different altitudes. *J. Fac. Sci. Shinshu University,* **10** (1), 27-34.

Shibata, O., Kinoshita, T. and Arai, T. (1975) Net production in several mature plants grown at different altitudes. *J. Fac. Sci. Shinshu University,* **10** (1), 35-39.

Shiosaka, H. and Shibata, O. (1992) Altitudinal variation of some morphological characteristics of *Polygonum cuspidatum* Sieb. et Zucc. achenes and seedlings. *Japanese Journal of Ecology,* **42**, 159-165.

Shiosaka, H. and Shibata, O. (1993) Morphological changes in *Polygonum cuspidatum* Sieb. et Zucc. reciprocally transplanted among different altitudes. *Japanese Journal of Ecology,* **43**, 31-37.

Sukopp, H. (1998) Anthropogenic plant migrations in Central Europe. In: Starfinger, U., Edwards, K., Kowarik, I. and Williamson, M. (eds), *Plant Invasions: Ecological Mechanisms and Human Responses.* Backhuys, Leiden. pp. 43-56.

Sukopp, H. and Starfinger, U. (1995) *Reynoutria sachalinensis* in Europe and the Far East: a comparison of the species ecology in its native and adventive distribution range. In: Pysek, P., Prach, K., Rejmánek, M. and Wade, P.M. (eds), *Plant Invasions: General Aspects and Special Problems.* SPB Academic Publishing, Amsterdam. pp. 151-159.

Sukopp, H. and Sukopp, U. (1988) *Reynoutria japonica* Houtt. in Japan und in Europa. *Veröff. Geobot. Inst. ETH*, **98**, 354-372.

Sukopp, H. and Schick, B. (1991) Zur biologie neophytischer *Reynoutria*-Arten in Mitteleuropa. I. Uber Floral- und Extrafloralnektarien. *Verh. Bot. Ver. Berlin Brandenburg*, **124**, 31-42.

Sukopp, H. and Schick, B. (1992) Zur biologie neophytischer *Reynoutria*-Arten in Mittleuropa. III. Morphometrie der Laubblatter. *Sonderdruck aus Natur und Landschaft*, **67** (10), 503-505. Kohlhammer, Stuttgart.

Sukopp, H. and Schick, B. (1993) Zur biologie neophytischer *Reynoutria*-Arten in Mitteleuropa: II. Morphometrie der Sprofssysteme. Festschrift Zoller. *Dissertationes Botanicae*, **196**, 163-174.

Suzuki, J.-I. (1994) Shoot growth dynamics and the mode of competition of two rhizomatous *Polygonum* species in the alpine meadow of Mt. Fuji. *Folia Geobot. Phytotax. Praha*, **29**, 203-216.

Suzuki, J.-I. (1994) Growth dynamics of shoot height and foliage structure of a rhizomatous perennial herb, *Polygonum cuspidatum*. *Annals of Botany*, **73**, 629-638.

Suzuki, N. (1988) Life history characteristics and resource utilization in the Chrysomelid species associated with Rumex plants. *Chrysomela*, **18**, 78.

Suzuki, N. (1994) Increased reproductive allocation of *Rumex japonicus* caused by leaf clipping. *Plant Species Biol.*, **9**, 31-35.

Tang, Y., Washitani, I. and Iwaki, H. (1992) Seasonal variations of microsite light availability within a *Miscanthus sinensis* canopy. *Ecological Research*, **7**, 97-106.

Tateno, M. (1988) Growth and turnover of microbial biomass during the decomposition of organic matter (*Polygonum cuspidatum*) in vitro. *Ecological Research*, **3**, 267-278.

Tateno, M. and Hirose, T. (1987) Nitrification and nitrogen accumulation in the early stages of primary succession on Mt. Fuji. *Ecological Research*, **2**, 113-120.

Tezuka, Y. (1961) Development of vegetation in relation to soil formation in the volcanic island of Oshima Izu, Japan. *Japanese Journal of Botany*, **17** (3), 371-402.

Trinajstic, I., Franjic, J. and Kajiba, D. (1994) Contribution to the knowledge of the spreading of the taxon *Reynoutria japonica* Houtt. (Polygonaceae) in Croatia. *Acta Botanica Croatica*, **53**, 145-149.

Tsuchida, K. (1973) The grassland vegetation in the subalpine zone of the Utsukushigahara Heights. Central Japan. *Japanese Journal of Ecology*, **23** (1), 33-43.

Tsuyuzaki, S. and Del Moral, R. (1994) Canonical correspondence analysis of early volcanic succession on Mt. Usu, Japan. *Ecological Research*, **9**, 143-150.

Tutin, T.G., Heywood, V.H., Burges, N.A., Valentine, D.H., Walters, S.M. and Webb, D.A. (eds), (1964) *Flora Europaea*. Cambridge University Press, London.

Welsh Development Agency (1994) *Model Tender Specifications for the Eradication of Japanese Knotweed*. Welsh Development Agency, Cardiff.

Welsh Development Agency (1998) *The Control of Japanese Knotweed in Construction and Landscape Contracts: Model Specification*. Welsh Development Agency, Cardiff.

Welsh Development Agency (1998) *The Eradication of Japanese Knotweed: Model Tender Document*. Welsh Development Agency, Cardiff.

Wolf, F.T. (1971) The growth rate of *Polygonum cuspidatum*. *Journal of the Tennessee Academy of Science*, **46**, 80.

Yamanishi, H. and Shibata, O. (1987) Physiological and ecological studies in environmental adaptation of plants. III. Altitudinal variation in some characters of cytochrome oxidase isozymes in *Polygonum cuspidatum* Sieb. et Zucc. *J. Fac. Sci. Shinshu University*. **22** (2), 75-82.

Young, R.G., Balogh, R.A., Sitler, T.R. and Aharrah, E.C. (1982) An investigation of Japanese Fleeceflower (*Polygonum cuspidatum*) planted on strip mines in Clarion and Venango Counties, Pennsylvania. *Symposium on Surface Mining Hydrology, Sedimentology and Reclamation*, University of Kentucky. pp. 143-152.

8.6 HISTORICAL ASPECTS OF JAPANESE KNOTWEED

Anon (1851) New plants, their portraits and biographies: spear-pointed-leaved *Polygonum*. *The Cottage Gardener*, **5**, 287-288.

Anon (1879) The knotweeds (Polygonums). *The Garden*, **XVI**. p.452.

Anon (1904) *Polygonum cuspidatum*. *Journal of the Royal Horticultural Society*, **29**, 180.

Bailey, L.H. (1927) *The Standard Cyclopedia of Horticulture*. Macmillan, London.

Baker, H.G. (1965) Characteristics and modes of origin of weeds. In: Baker, H.G. and Stebbins, G.L. (eds), *The Genetics of Colonizing Species*. Academic Press, New York. pp. 147-172.

Conolly, A.P. (1977) The distribution and history in the British Isles of some alien species of *Polygonum* and *Reynoutria*. *Watsonia*, **11**, 291-311.

Elton, C.S. (1958) *The Ecology of Invasions by Animals and Plants*. Chapman and Hall, London.

Groenland, J. (1858) *Polygonum cuspidatum. Revue Horticole*, 630-639.

Hooker, J.D. (1880) *Polygonum cuspidatum. Botanical Magazine*, **106**, Tab 6503.

Hooker, J.D. (1880) *Polygonum compactum. Botanical Magazine*, **106**, Tab 6476.

Hooker, J.D. (1881) *Polygonum sachalinense. Botanical Magazine*, **107**, Tab 6540.

Lindley, J. and Paxton, J. (1850-51) Gleanings and original memoranda. *Flower Garden*. 137-138.

MacLean (1978) Von Siebold and the importation of Japanese plants into Europe via the Netherlands. *Japanese Studies in the History of Science*, **17**, 43-79.

Makino (1901) *Polygonum Reynoutria* (Houtt.) Makino nom. nov. *Bot. Mag. Tokyo*, **15**, 84.

Patterson, D. T . (1976) The history and distribution of five exotic weeds in N. Carolina. *Castanea*, **41** (2), 177-180.

Storrie, J. (1886) *The Flora of Cardiff*. Cardiff.

8.7 IDENTIFICATION OF JAPANESE KNOTWEED

Alberternst, B., Bauer, M., Boecker, R. and Konold, W. (1995) *Reynoutria* species in Baden-Wuerttemberg: Keys for the determination and their distribution along fresh waters. *Floristische Rundbriefe*, **29**, 113-124.

Ellis, G. (1989) A Japanese knotweed by any other name. *BSBI News*, **53**, 34.

Lousley, J.E. and Kent, D.H. (1981) *Docks and Knotweeds of the British Isles*. BSBI.

Makino (1901) *Polygonum Reynoutria* (Houtt.) Makino nom. nov. *Bot. Mag. Tokyo*, **15**, 84.

Meyer, F.G. (1970) Mountain fleece flower re-identified (Polygonum cuspidatum 'Crimson Beauty'). *Am. Hortic.*, **49** (3), 132.

Ohba, T. (1975) Uber die *Polygonum cuspidatum* var. *terminale – Carex doenitsii* var. *okuboi* – Ass. ass. nov. mit einer Bemerkung ₂ber der Ursprung der speziellen Flora der Izu-Inseln Japans. *Bull. Kanagawa Pref. Mus.*, **8**, 91-106.

Ohwi, J. (1965) *Flora of Japan*. Smithsonian Institution, Washington, DC.

Rich, T.C.G. and Rich, M.D.B. (1988) *Plant Crib*. Botanical Society of the British Isles, London.

Stace, C. (1991) *New Flora of the British Isles*. Cambridge University Press, Cambridge.

8.8 PROBLEMS ASSOCIATED WITH JAPANESE KNOTWEED

Beerling, D.J. (1991a) The testing of cellular concrete revetment blocks resistant to growths of *Reynoutria japonica* Houtt. (Japanese knotweed). *Water Research*, **25**, 495-498.

Beerling, D.J. (1995) General aspects of plant invasions: an overview. In: Pysek, P., Prach, K., Rejmánek, M. and Wade, P.M. (eds), *Plant Invasions: General Aspects and Special Problems*. SPB Academic Publishing, Amsterdam. pp. 237-247.

Cooke, A.S. (1988) Japanese knotweed: its status as a pest and its control in nature conservation areas. Unpublished Report, English Nature, Peterborough.

Crawley, M.J. (1989) Invaders. *Plants Today*, September – October 1989, 152-158.

Dawson, H.J. (1986) *Petrorhagia nanteulii* (Burnat) P.W. Ball & Heywood in Mid-Glamorgan. *Watsonia*, **16** (2), 174-175.

Department of the Environment (1991) *Waste Management – The Duty of Care, a Code of Practice. Environmental Protection Act 1990*. HMSO, London.

Dowsett-Lemaire, F. (1981) Eco-ethological aspects of breeding in the marsh warbler, *Acrocephalus Palustris. Rev. Ecol. (Terre et Vie)*, **35**, 451-485.

Environment Agency (1996) Guidance for the *Control of Invasive Plants near Watercourses*. Environment Agency, Bristol.

Environment Agency and Cornwall County Council (1998) *Japanese Knotweed. How to Control It and Prevent Its Spread*. Environment Agency, Bodmin.

Ezaki, T. (1984) Studies on the conservation of the face of slopes of forest roads. *Bulletin of the Ehime University Forest*, 21, 11-16.

Gilbert, O.L. (1994) Japanese knotweed – what problem? *Urban Wildlife News*, **11**, 1.

Wildlife and Countryside Act 1981. HMSO, London.

8.9 TAXONOMY OF JAPANESE KNOTWEED AND RELATED SPECIES

Bailey, J.P. (1988) Putative *Reynoutria japonica* Houtt. x *Fallopia baldschuanica* (Regel) Holub hybrids discovered in Britain. *Watsonia*, **17**, 163-181.

Bailey, J.P. (1989) Cytology and breeding behaviour of giant alien *Polygonum* species in Britain. Unpublished Ph.D. thesis, University of Leicester, UK.

Bailey, J.P. (1990) Breeding behaviour and seed production in alien giant knotweed in the British Isles. *Conference of Industrial Ecology Group of the British Ecological Society*. Cardiff. pp. 121-129.

Bailey, J.P. (1992) The Haringey knotweed. *Urban Nature Magazine*, Autumn 1992, 50-51.

Bailey, J.P. (1994) The reproductive biology and fertility of *Fallopia japonica* (Japanese knotweed) and its hybrids in the British Isles. In: Waal, L.C. de, Child, L.E., Wade, P.M. and Brock, J.H. (eds), *Ecology and Management of Invasive Riparian Plants*. Wiley, Chichester. pp. 141-158.

Bailey, J.P. (1997) The Japanese knotweed invasion of Europe: the potential for further evolution in non-native regions. In: National Institute of Agro-Environmental Sciences, *Proceedings International Workshop on Biological Invasions of Ecosystems by Pests and Beneficial Organisms*, February 25-27, 1997, Tsukuba, Japan. pp. 32-47.

Bailey, J.P. and Conolly, A. P. (1984) A putative *Reynoutria* x *Fallopia* hybrid from Wales. *Watsonia*, **15**, 162-163.

Bailey, J.P. and Conolly, A.P. (1985) Chromosome numbers of some alien *Reynoutria* species in the British Isles. *Watsonia*, **15**, 270-271.

Bailey, J.P. and Conolly, A.P. (1991) Alien species of *Polygonum* and *Reynoutria* in Cornwall 1989-1990. *Botanical Cornwall Newsletter*, **5**, 33-46.

Bailey, J.P. and Stace, C.A. (1992) Chromosome number, morphology, pairing, and DNA values of species and hybrids in the genus *Fallopia* (Polygonaceae). *Plant Systematics and Evolution*, **180**, 29-52.

Bailey, J.P., Child, L.E. and Conolly, A.P. (1996) A survey of the distribution of *Fallopia* x *bohemica* (Chrtek and Chrtková) J. Bailey (Polygonaceae) in the British Isles. *Watsonia*, **21**, 187-198.

Bailey, J.P., Child, L.E. and Wade, P.M. (1995) Assessment of the genetic variation and spread of British populations of *Fallopia japonica* and its hybrid *Fallopia* x *bohemica*. In: Pysek, P., Prach, K., Rejmánek, M. and Wade, P.M. (eds), *Plant Invasions: General Aspects and Special Problems*. SPB Academic Publishing, Amsterdam. pp. 141-150.

Chrtek, J. and Chrtkov·, A. (1983) *Reynoutria* x *bohemica*, Novy Krieznec Z celedi Rdesnovitych. *J. Nat. Mus. Praha Hist. Nat.*, **152**, 120 (in Czech).

Coombes, A.J. (1985) *The Collingridge Dictionary of Plant Names*. Collingridge, London.

Doida, Y. (1960) Cytological studies in Polygonum and related genera. *Bot. Mag. Tokyo*, **37**, 337-340.

Lousley, J.E. and Kent, D.H. (1981) *Docks and Knotweeds of the British Isles.* Botanical Society of the British Isles, London.

Rich, T.C.G. and Rich, M.D.B. (1988) *Plant Crib*. Botanical Society of the British Isles, London.

Ronse Decraene, L-P. and Akeroyd, J.R. (1988) Generic limits in *Polygonum* and related genera (Polygonaceae) on the basis of floral characters. *Botanical Journal of the Linnean Society*, **98**, 321-371.

Stace, C. (1991) *New Flora of the British Isles*. Cambridge University Press, Cambridge.

Steffey, J. (1980) The buckwheat family (Polygonaceae, *Polygonum cuspidatum*). *American Horticulturist*, **59** (7), 10-11.

Sukopp, H. and Starfinger, U. (1995) *Reynoutria sachalinensis* in Europe and the Far East: A comparison of the species ecology in its native and adventive distribution range. In: Pysek, P., Prach, K., Rejmánek, M. and Wade, P.M. (eds), *Plant Invasions: General Aspects and Special Problems*. SPB Academic Publishing, Amsterdam, The Netherlands. pp. 151-159.

Sukopp, H. and Sukopp, U. (1988) *Reynoutria japonica* Houtt. in Japan und in Europa. *Veröff. Geobot. Inst.* ETH, **98**, 354-372.

Weber, E.F. (1997) The alien flora of Europe: a taxonomic and biogeographic review. *Journal of Vegetation Science*, **8**, 565-572.

8.10 OTHER USEFUL RESOURCES

Anon (1982) The relation between flowers, seeds and fruits. In: *Principles of Dispersal in Higher Plants*. Springer-Verlag, Berlin.

Ashton, P.J. and Mitchell, D.S. (1989) Aquatic plants: patterns and modes of invasion, attributes of invading species and assessment of control programmes. In: Drake, J.A., Mooney, H.A., di Castri, F., Groves, R.H., Kruger, F.J., Rejmánek, M. and Williamson, M. (eds), *Biological Invasions: a Global Perspective*. SCOPE 37. Wiley, Chichester.

Baker, H.G. (1986) Patterns of plant invasion in North America. In: Mooney, H.A., Drake, J.A. (eds), *Ecology of Biological Invasions of North America and Hawaii*. Springer-Verlag, New York. 44-57.

Barker, G. (1996) Alien and native – drawing the line. *Ecos*, **17**.

Bazzaz, F.A. (1986) Life history of colonising plants: some demographic, genetic and physiological features. In: Mooney, H.A. and Drake, J.A. (eds), *Ecology of Biological Invasions of North America and Hawaii*. Springer-Verlag, New York. 96-110.

Beerling, D.J., and Perrins, J.M. (1993) Biological Flora of the British Isles *Impatiens glandulifera* Royle (*Impatiens roylei* Walp.). *Journal of Ecology*, **81**, 367-382.

Binggeli, P. (1994) The misuse of terminology and anthropometric concepts in the description of introduced species. *Bull. Brit. Ecol. Soc.*, **25**, 10-13.

Bradshaw, T.K. and Muller, B. (1998) Impacts of rapid urban growth on farmland conversion: Applications of new regional landuse policy models and geographical information systems. *Rural Sociology*, **63**, 1-25.

Briggs, D. and Mounsey, H. (1989) Integrating land resource data into a European geographical information system: practicalities & problems. *Applied Geography*, **9**, 5-20.

Brown, N.J., Swetnam, R.D., Treweek, J.R., Mountford, J.O., Caldow, R.W.G., Manchester, S.J., Stamp, T.R., Gowning, D.J.G., Solman, D.R. and Armstrong, A.C. (1998) Issues in Geographical Information Systems development: adapting to research and policy needs for management of wet grasslands in an Evironmentally Sensitive Area. *International Journal of Geographical Information Science*, **12**, 465-478.

Caffrey, J.M. (1994) Spread and management of *Heracleum mantegazzianum* (giant hogweed) along Irish river corridors. In: Waal, L.C. de, Child, L.E., Wade P.M. and Brock, J.H. (eds), *Ecology and Management of Invasive Riparian Plants*. Wiley, Chichester. pp. 67-76.

Centre for Aquatic Plant Management (1997) Control of floating pennywort (*Hydrocotyle ranunculoides*). Information Sheet 20, Centre for Aquatic Plant Management, Reading.

Child, L.E. and Spencer-Jones, D. (1995) Treatment of *Crassula helmsii* – a case study. In: Pysek, P., Prach, K., Rejmánek, M. and Wade, P.M. (eds), *Plant invasions: General Aspects and Special Problems*. SPB Academic Publishing, Amsterdam. pp. 195-202.

Chorley, R. and Buxton, R. (1991) The government setting of GIS in the United Kingdom. In: Maguire, D.J., Goodchild, M.F. and Rhind, D.W. (eds), *Geographical Information Systems: Principles and Applications*. Longman, Harlow. pp. 67-79.

Crawley, M.J. (1989) Chance and timing in biological invasions. In: Drake, J.A., Mooney, H.A., di Castri, F., Groves, R.H., Kruger, F.J., Rejmánek, M. and Williamson, M. (eds), *Biological Invasions: a Global Perspective*. SCOPE 37. Wiley, Chichester. pp. 407-424.

Crawley, M.J. (1987) What makes a community invasible? In: Gray, A.J., Crawley, M.J. and Edwards, P.J. (eds), Colonization, succession and stability. *Symposia of the British Ecological Society*, **26**, Blackwell Scientific, Oxford. pp 429-454.

Cronk, Q.C.B. and Fuller, J.L. (1995) *Plant Invaders*. Chapman and Hall, London.

Davis, Langdon and Everest (1992) *Spon's Architects' and Builders' Price Book*. E. and F.N. Spon, London.

Dawson, F.H. (1994) Spread of *Crassula helmsii* in Britain. In: Waal, L.C. de, Child, L.E., Wade, P.M. and Brock, J.H. (eds), *Ecology and Management of Invasive Riparian Plants*. Wiley, Chichester. pp. 1-14.

Dawson F.H. and Henville, P. (1991) *An investigation into the Control of Crassula helmsii by Herbicidal Chemicals*. Final Report to Nature Conservancy Council. pp. 107.

di Castri, F., Hansen, A.J. and Debussche, M. (eds) (1990) *Biological Invasions in Europe and the Mediterranean Basin*. Kluwer, Dordrecht.

Drake, J.A., Mooney, H.A., di Castri, F., Groves, R.H., Kruger, F.J., Rejmánek, M. and Williamson, M. (eds) (1989) *Biological Invasions: a Global Perspective*. SCOPE 37. Wiley, Chichester.

Duncan, K.W. (1997) A case study in *Tamarix ramosissima* control: Spring Lake, New Mexico. In : Brock, J.H., Wade, P.M., Pysek, P. and Green, D. (eds), *Plant Invasions: Studies from North America and Europe*, Backhuys, Leiden. Ppp. 115-121.

Enomoto, T. (1997) Naturalised weeds from foreign country into Japan. In: National Institute of Agro-Environmental Sciences, *Proceedings International Workshop on Biological Invasions of Ecosystems by Pests and Beneficial Organisms*, February 25-27, 1997, Tsukuba, Japan. 2-15.

Eser, U. (1998) Assessment of plant invasions: theoretical and philosophical fundamentals. In: Starfinger, U., Edwards, K., Kowarik, I. and Williamson, M. (eds), *Plant Invasions: Ecological Mechanisms and Human Responses*. Backhuys, Leiden. pp. 95-107.

Everitt, J.H., Escobar, D.E. and Davis, M.R. (1995) Using remote sensing for detecting and mapping noxious plants. *Weed Abstracts*, **44**, 639 - 649. Centre for Agricultural and Biosciences (CAB) International.

Fox, M.D. and Fox, B.J. (1986) The susceptibility of natural communities to invasion. In: Groves, R.H. and Burdon, J.J. (eds), *Ecology of Biological Invasions*. Cambridge University Press. Sydney. pp. 57-66.

Goodchild, M.F. (1991) The Technological setting of GIS. In: Maguire, D.J., Goodchild, M.F. and Rhind, D.W. (eds), *Geographical Information Systems: Principles and Applications*. Longman, Harlow. pp. 45-54.

Gritten, R.H. (1995) *Rhododendron ponticum* and some other invasive plants in the Snowdonia National Park. In: Pysek, P., Prach, K., Rejm·nek, M. and Wade, P.M. (eds), *Plant Invasions: General Aspects and Special Problems*. SPB Academic Publishing, Amsterdam. pp. 213-219.

Hendriks, P.H.J. (1998) Information strategies for Geographical Information Systems. *International Journal of Geographical Information Science*, **12**, 621-639.

Hengeveld, R. (1989) *Dynamics of Biological Invasions*. Chapman and Hall, London.

Heywood, I. (1990) Monitoring for change. A Canadian perspective on the environmental role for GIS. *Mapping Awareness*, **4**, 24-26.

Heywood, V.H. (1989) Patterns, extents and modes of invasions by terrestrial plants. In: Drake, J.A., Mooney, H.A., di Castri, F., Groves, R.H., Kruger, F.J., Rejmánek, M. and Williamson, M. (eds), *Biological Invasions: a Global Perspective*. SCOPE 37. Wiley, Chichester. pp. 31-60.

Hobbs, R.J. and Humphries, S.E. (1994) An integrated approach to the ecology and management of plant invasions. *Conservation Biology*, **9**, 761-770.

Holdgate, M.W. (1987) Summary and conclusions: characteristics and consequences of biological invasions. In: Kornberg, H. and Williamson, M.H. (eds), *Quantitative Aspects of the Ecology of Biological Invasions*. The Royal Society, London.

Janes, R.A. (1995) The Biology and Control of *Azolla filiculoides Lam. and Lemna minuta Kunth*. Unpublished Ph.D. thesis, University of Liverpool, UK.

Kornas, J. (1990) Plant invasions in Central Europe: historical and ecological aspects. In: di Castri, F., Hansen, A.J. and Debussche, M. (eds) (1990) *Biological Invasions in Europe and the Mediterranean Basin*. Kluwer, Dordrecht. pp. 19-36.

Kowarik, I. (1990) Some responses of flora and vegetation to urbanization in central Europe. In: Sukopp, H., Henjy, S. and Kowarik, I. (eds), *Urban Ecology*. SPB Academic Publishers, The Hague. pp. 45-74.

Kowarik, I. (1995a) Time lags in biological invasions with regard to the success and failure of alien species. In: Pysek, P., Prach, K., Rejmánek, M. and Wade, P.M. (eds), *Plant Invasions: General Aspects and Special Problems*. SPB Academic Publishing, Amsterdam. pp. 15-38.

Kowarik, I. (1995b) On the role of alien species in urban flora and vegetation. In: Pysek, P., Prach, K., Rejmánek, M. and Wade, P.M. (eds), *Plant Invasions: General Aspects and Special Problems*. SPB Academic Publishing, Amsterdam. pp. 85-103.

Lodwick, A. and Cushnie, J. (1990) A GIS pilot study in Berkshire. *Mapping Awareness*, **4**, 39-42.

Mack, R.N. (1986) Alien plant invasion into the intermountain west: a case history. In: Mooney, H.A. and Drake, J.A. (eds), *Ecology of Biological Invasions of North America and Hawaii.* Springer-Verlag, New York. pp. 91-211.

Maguire, D.J. (1991) An overview and definition of GIS. In: Maguire, D.J., Goodchild, M.F. and Rhind, D.W. (eds), *Geographical Information Systems: Principles and Applications.* Longman, Harlow. pp. 9-20.

Markham, R., Rix, D. and Bentley, R. (1990) GIS at Plymouth City Council – coming to terms with reality. *Mapping Awareness*, **4**, 10-13.

Navas, M.L. (1991) Using plant population biology in weed research: a strategy to improve weed management. *Weed Research*, **31**, 171-179.

Noble, I.R. (1989) Attributes of invaders and the invading process: terrestrial and vascular plants. In: Drake, J.A., Mooney, H.A., di Castri, F., Groves, R.H., Kruger, F.J., Rejmánek, M. and Williamson, M. (eds), *Biological Invasions: a Global Perspective.* SCOPE 37. Wiley, Chichester. pp. 301-314.

Orians, G.H. (1986) Site characteristics favoring invasions. In: Mooney, H.A. and Drake, J.A. (eds), *Ecology of Biological Invasions of North America and Hawaii.* Springer-Verlag, New York. pp.133-148.

Parrott, R. and Stutz, F.P. (1991) Urban GIS applications. In: Maguire, D.J., Goodchild, M.F. and Rhind, D.W. (eds), *Geographical Information Systems: Principles and Applications.* Longman, Harlow. pp. 247-260.

Perrins, J., Williamson, M. and Fitter, A. (1992) A survey of differing views of weed classification: implications for regulation of introductions. *Biological Conservation*, **60**, 47-56.

Preston, C.D. (1986) An additional criterion for assessing native status. *Watsonia*, **16**, 83.

Pysek, P. (1991) *Heracleum mantegazzianum* in the Czech Republic: the dynamics of spreading from the historical perspective. *Folia Geobot. Phytotax.*, **26**, 439-454.

Pysek, P. (1995) On the terminology used in plant invasion studies. In: Pysek, P., Prach, K., Rejmánek, M. and Wade, P.M. (eds), *Plant Invasions: General Aspects and Special Problems.* SPB Academic Publishing, Amsterdam. pp. 71-81.

Pysek, P. (1997) Clonality and plant invasions: can a trait make a difference? In: de Kroon, H. and van Groenendael, J. (eds), *The Ecology and Evolution of Clonal Plants.* Backhuys, Leiden. pp. 405-427.

Pysek, P. and Prach, K. (1993) Plant invasions and the role of riparian habitats: a comparison of four species alien to central Europe. *Journal of Biogeography*, **20**, 413-420.

Pysek, P., Prach, K. and Smilauer, P. (1995) Relating invasion success to plant traits: an analysis of the Czech alien flora. In: Pysek, P., Prach, K., Rejmánek, M. and Wade, P.M. (eds), *Plant Invasions: General Aspects and Special Problems.* SPB Academic Publishing, Amsterdam. pp. 39-60.

Rejmánek, M. (1989) Invasibility of plant communities. In: Drake, J.A., Mooney, H.A., di Castri, F., Groves, R.H., Kruger, F.J., Rejmánek, M. and Williamson, M., (eds), *Biological Invasions: a Global Perspective.* SCOPE 37. Wiley, Chichester. pp. 369-388.

Rejmánek, M. (1995) What makes a species invasive? In: Pysek, P., Prach, K., Rejmánek, M. and Wade, P.M. (eds), *Plant Invasions: General Aspects and Special Problems.* SPB Academic Publishing, Amsterdam. pp. 3-13.

Roy, J. (1990) In search of the characteristics of plant invaders. In: di Castri, F., Hansen, A.J. and Debussche, M. (eds), *Biological Invasions in Europe and the Mediterranean Basin.* Kluwer, Dordrecht. pp. 335-352.

Roy, J., Navas, M.L. and Sonié, L. (1991) Invasion by annual brome grasses: a case study challenging the homocline approach to invasions. In: Groves, R.H. and di Castri, F. (eds), *Biogeography of Mediterranean Invasions.* Cambridge University Press, Cambridge. pp. 207-224.

Schmieder, K. (1995) Application of Geographical Information Systems (GIS) in lake monitoring with submersed macrophytes at Lake Constance – Conception and purposes. *Acta Botanica Gallica,* **142**, 551-554.

Sheley, R.L., Mullin, B.H. and Fay, P.K. (1995) Managing riparian weeds. *Rangelands,* **17**, 154-157.

Sui, D.Z. (1998) GIS based urban modelling: practices, problems and prospects. *International Journal of Geographical Information Science,* **12**, 651-671.

Thomson, A.G., Radford, G.L., Norris, D.A. and Good, J.E.G. (1993) Factors affecting the distribution and spread of *Rhododendron* in North Wales. *Journal of Environmental Management,* **39**, 199-212.

Tiley, G.E.D. and Philp, B. (1997) Observations on flowering and seed production in *Heracleum mantegazzianum* in relation to control. In: Brock, J.H., Wade, P.M., Pysek, P. and Green, D. (eds), *Plant Invasions: Studies from North America and Europe.* Backhuys, Leiden. pp.123-127.

Tiley, G.E.D., Dodd, F.S. and Wade, P.M. (1996) Biological Flora of the British Isles: *Heracleum mantegazzianum* Sommier and Levier. *Journal of Ecology,* **84**, 297-319.

Vitousek, P.M. (1986) Biological invasions and ecosystem propoerties: can species make a difference? In: Mooney, H.A. and Drake, J.A. (eds), *Ecology of Biological Invasions of North America and Hawaii.* Springer-Verlag, New York. pp. 164-174.

Vogt-Andersen, U. (1995) Dispersal strategies of alien and native species. In: Pysek, P., Prach, K., Rejmánek, M. and Wade, P.M. (eds), *Plant Invasions: General Aspects and Special Problems*. SPB Academic Publishing, Amsterdam. pp. 61-70.

Wade, P.M., Darby, E.J., Courtney, A.D. and Caffrey, J.M. (1997) *Heracleum mantegazzianum*: a problem for river managers in the Republic of Ireland and the United Kingdom. In: Brock, J.H., Wade, P.M., Pysek, P. and Green, D. (eds), *Plant Invasions: Studies from North America and Europe*. Backhuys, Leiden. pp. 139-151.

Webb, D.A. (1985) What are the criteria for presuming native status? *Watsonia*, **15**, 231-236.

Weston, J. (1995) Planning is paramount: GIS in district councils. *Mapping Awarenes*, **10**, 32-35.

Williamson, M. (1993) Invaders, weeds and the risk of genetically manipulated organisms. *Experientia*, **49**, 219-224.

Williamson, M. (1996) *Biological Invasions*. Chapman and Hall, London.

Williamson, M. (1998) Measuring the impact of plant invaders in Britain. In: Starfinger, U., Edwards, K., Kowarik, I. and Williamson, M. (eds), *Plant Invasions: Ecological Mechanisms and Human Responses*. Backhuys, Leiden. pp. 57-68.

Williamson, M. and Brown, K.C. (1986) The analysis and modelling of British invasions. *Phil. Trans. R. Soc. Lond.* **B314**, 505-522.

Williamson, M. and Fitter, A. (1996) The characters of succesful invaders. *Biological Conservation*, **78**, 163-170.

World Resources Institute (1998) Web site http://www.wri.org

Worrall, L. and Bond, D. (1997) Geographical Information Systems, spatial analysis and public policy: the British experience. *International Statistical Review*, **65**, 365-379.

Zhu, X.A., Healey, R.G. and Aspinall, R.J. (1998) Knowledge based systems approach to design of spatial decision support systems for environmental management. *Environmental Management*, **22**, 35-48.